SpringerBriefs in Earth Sciences

W0193125

More information about this series at http://www.springer.com/series/8897

Jerry R. Miller · Gail Mackin
Suzanne M. Orbock Miller

Application of Geochemical Tracers to Fluvial Sediment

 Springer

Jerry R. Miller
Department of Geosciences and Natural
  Resources
Western Carolina University
Cullowhee, NC
USA

Suzanne M. Orbock Miller
Haywood County Schools
Waynesville, NC
USA

Gail Mackin
Department of Mathematics
Northern Kentucky University
Highland Heights, KY
USA

ISSN 2191-5369   ISSN 2191-5377 (electronic)
SpringerBriefs in Earth Sciences
ISBN 978-3-319-13220-4   ISBN 978-3-319-13221-1   (eBook)
DOI 10.1007/978-3-319-13221-1

Library of Congress Control Number: 2014956374

Springer Cham Heidelberg New York Dordrecht London

Printed on acid-free paper

Springer International Publishing AG Switzerland is part of Springer Science+Business Media (www.springer.com)

To Dale (Dusty) F. Ritter
(1932–2012)
Teacher, Mentor, Colleague, and Friend

Jerry R. Miller
Suzanne M. Orbock Miller

To Timothy J. Hodges
For His Constant Support and Encouragement

Gail Mackin

Acknowledgments

We want to extend our sincere thanks to a number of individuals who graciously devoted their time and effort to reviewing the chapters in this book in a highly constrained timeframe. They include Drs. Dru Germanoski, Timothy Hodges, Karen Hudson-Edwards, Mark Lord, and Mark Taylor. Their in-depth comments and suggestions greatly improved both the content and the presentation of the covered materials. Thanks also go to our friends and colleagues with whom we have spent countless hours discussing the science of geochemical fingerprinting. It was through these discussions that we were encouraged to undertake this endeavor with the hopes of providing insights to other scientists, regulators, and resource managers on the benefits of the approach.

Jerry R. Miller
Gail Mackin
Suzanne M. Orbock Miller

Contents

Chapter 1
Introduction

1.1 Tracers, Fingerprints, and Riverine Sediments

Tuero Chico is a small village located along the Rio Pilcomayo of southern Bolivia. Soils associated with its farmed floodplains possess Pb concentrations that exceed recommended guidelines for agricultural use. The elevated levels of Pb raise a number of important questions: Is the Pb derived from upstream mining of the Potosi precious metal-polymetallic tin deposits, or waste products disposed of in the river from the City of Potosi? Perhaps it is natural, being derived from local mineralized rocks that underlie the catchment? Or, could the Pb come from a combination of all three sources? If it is from multiple sources, how much comes from each source? And, how far downstream does the Pb from a specific source impact sediment and water quality? These and other complex physical and biogeochemical questions are increasingly being addressed using environmental tracers. In this book we examine the past, current, and future use of environmental tracers to assess the provenance, movement, and ultimate fate of sediment within river systems, particularly sediments contaminated by chemical substances that have the potential to degrade aquatic ecosystems and/or human health. The term *tracer* has been defined in different ways depending on the media (e.g., air, ice, snow, ground- or surface waters) to which it is applied. For our purposes, a *tracer* is defined as a unique sediment-associated parameter or set of parameters that is distinct from other sediments in the catchment, and can therefore be used to track the movement and cycling of specific sediments from their point of origin to their ultimate point of deposition. The term 'tracer' is often defined and used synonymously with *fingerprint*. However, when applied to river (fluvial) systems, a fingerprint is most commonly associated with a specific type of analyses (fingerprinting studies) in which multiple parameters are used to distinguish between sediments from diffuse (non-point) sources to quantify the provenance of the sediment found in a river or riverine deposit.

The use of fingerprinting and tracing methods to assess the dynamics of sediment generation, transport and storage has a long history in both fluvial sedimentology and geomorphology, dating back to at least the early 20th century (e.g., Boswell 1933).

© The Author(s) 2015   1
J.R. Miller et al., *Application of Geochemical Tracers to Fluvial Sediment*,
SpringerBriefs in Earth Sciences, DOI 10.1007/978-3-319-13221-1_1

It was not until the middle of the 1900s, however, that the potential for tracer studies to provide meaningful data on sediment dynamics began to be appreciated (Walling et al. 2013). Early studies were primarily aimed at understanding particle entrainment thresholds and transport distances of large bed material clasts within short reaches of the channel and were based on what Black et al. (2007) calls 'particle tracking'. Essentially, particle tracking refers to (1) the practice of tagging individual clasts in some fashion so their movement can be documented, especially during storm events, or (2) the addition of exotic constituents to a mixture of sediment so that the movement of sediment of similar characteristics can be monitored. These studies initially relied on rather unsophisticated methods (e.g., painting of a particle surface), but have evolved so that particle tracking now includes such sophisticated technologies as inserting magnets or radio-transmitters into individual clasts of varying size, or incorporating Rare Earth Elements, magnetic constituents (e.g., magnetite), and other materials in the sediment to monitor their incipient motion and transport distances in near real-time (Parsons et al. 1993; Zhang et al. 2003; Kimoto et al. 2006; Mentler et al. 2009; Guzmán et al. 2010; Hu et al. 2011; Spencer et al. 2011). These techniques can also be used to assess such things as transport step lengths and rest periods for variously sized particles, and have been applied to other problems such as soil erosion rates and redistribution patterns on hillslopes.

The 1980s and 1990s saw an expansion of tracer research to address a number of additional aspects of the sediment system, including the origin and transport mechanisms of particles found in both consolidated (sedimentary) and unconsolidated deposits. Walling et al. (2013) point out that these studies differed from earlier particle tracking methods in three important ways. First, particle tracking as originally conducted required the addition of a tracer material which was costly to use over large areas; thus, the addition of a tracer was (and continues to be) restricted to short reaches of river channel or small soil plots. To circumvent this problem, investigators began to utilize natural characteristics of the sediment (e.g., its mineralogy, grain size, color, chemical composition, and magnetic properties) as a tracer, or utilize some pre-existing constituent within the sediment. With respect to the latter, tracers often consisted of anthropogenic constituents (e.g., ^{137}Cs from surficial nuclear bomb tests or trace metals from mining operations). Second, the use of natural and pre-existing tracers allowed the area of study to be greatly expanded from short river reaches or small soil plots to the landscape scale. From this larger scale perspective, the sediment system can be envisioned as an integrated sediment generation and dispersal network in which sediments are produced in upland areas and ultimately deposited downstream in a basin that acts as a long-term repository (Fig. 1.1). These zones of sediment production and deposition are connected by a drainage network that intermittently moves sediment, primarily during flood events, from source to sink (Schumm 1977; Weltje 2012). Tracers, at this scale, can be used to address aspects of the entire, and highly complex, sediment dispersal system over a variety of temporal scales. Third, fingerprinting and tracing methods began to focus upon the fine-grained sediment fraction, rather than the coarse-grained bed load (Walling et al. 2013). Interest in fine-sediments resulted from the fact that the excessive generation and transport of particulates $< \sim 2$ mm in size pose a direct threat

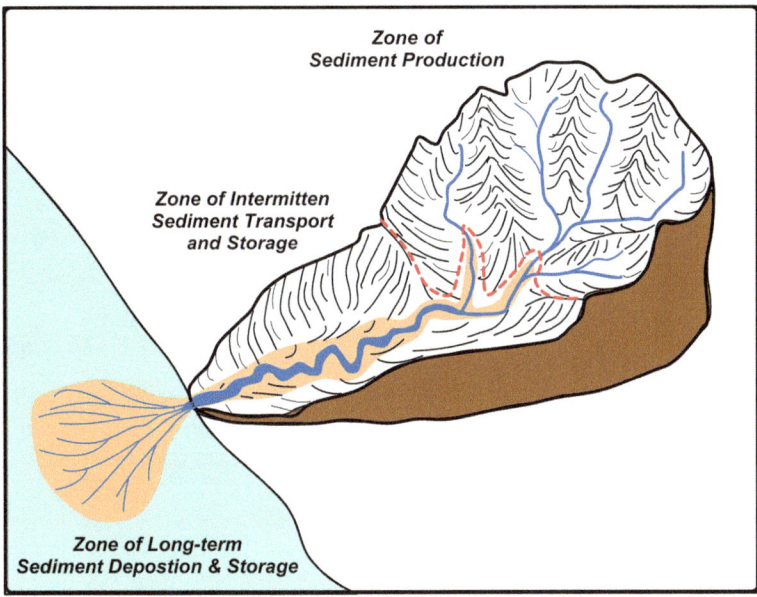

Fig. 1.1 Schematic diagram of riverine sediment-dispersal system (after Schumm 1977)

to ecosystem health. For example, the National Water Quality Inventory, a program in the U.S. developed to assess the current condition of the nations water resources, indicates that sediment is the second leading cause of river impairment (Fig. 1.2) (USEPA 2013). Moreover, anthropogenically derived sediment can result in rapid episodes of reservoir sedimentation, reduce reservoir storage capacity, impact water distribution systems, increase turbidity and reduce light penetration, degrade aquatic habitat, and lead to a loss in aesthetic quality of the riverine environment. The annual costs of human-induced sediment influx to rivers and streams have been estimated to range from 20 to 50 billion dollars in North America alone (Pimentel et al. 1995; Osterkamp 2004; Mukundan et al. 2012).

From a chemical perspective, fine-grained sediments, particularly those composed of clay minerals, Fe and Mn oxides and hydroxides, and organic matter are highly reactive (Horowitz 1991). Thus, sediment suspended within the water column and that forms the channel bed and banks, typically exhibit concentrations of hydrophobic contaminants that are orders of magnitude higher than those associated with the aqueous (dissolved) load. Gibbs (1977), for example, examined the concentration of selected metals (including Cu, Co, Cr, Fe, Mn, and Ni) associated with suspended sediment within the Yukon and Amazon River basins, two river systems characterized by different hydrologic regimes and geological terrains. He found that within both basins sediment-associated trace metal levels ranged from 6,000 to more than 10,000 times greater than their dissolved concentrations. As a result, trace metal transport was dominated by the particulate load (Fig. 1.3). Subsequent studies (e.g., Horowitz and Elrick 1988; Meybeck and Hemler 1989; Horowitz 1991) supported Gibbs'

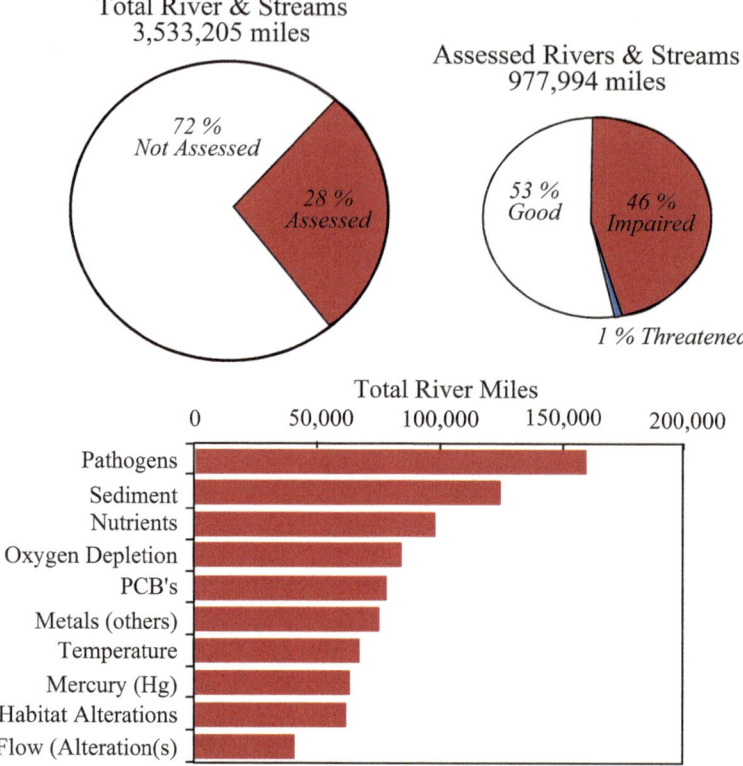

Fig. 1.2 Leading causes of river impairment in the U.S. as determined by the National Water Quality Inventory. Data reported for 2010–2012 (depending on state). Note that sediment is the second leading cause of impairment (data from USEPA 2013)

observed differences between dissolved and particulate concentrations. This led to the argument that within rivers exhibiting typically observed pH and Eh conditions more than 90 % of the trace metal load is transported as part of the sediment load (Table 1.1).

The potential for sediment and sediment-associated contaminants to negatively impact aquatic ecosystems has led to the general evolution in the application of environmental tracers from a state in which they were primarily used in academic studies to their use as a management and regulatory tool. While this evolution in tracer utilization has been slower with regards to rivers than it has been for, say, groundwater, it is likely to progress in the future. It is also likely to be closely linked to the developing field of *Environmental Forensics*. Haddad (2004) described Environmental Forensics as "that part of the Venn Diagram where environmental technical questions overlap legal issues". A more detailed definition, put forth by Wenning and Simmons (2000), is the "systematic examination of environmental information to determine sources of chemical contamination, the timing of releases to the environment, the spatial distribution of contamination, and the potential responsible party(ies)". It seems

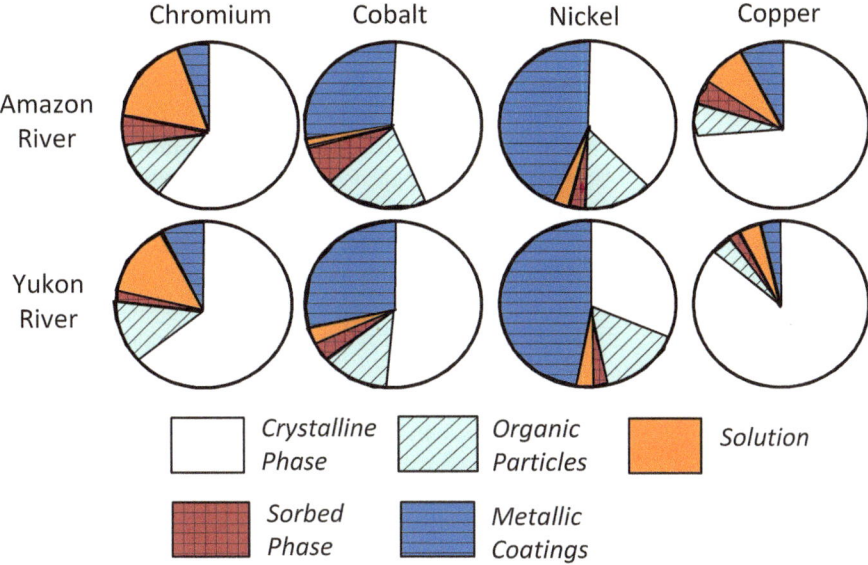

Fig. 1.3 Percentages of selected metals transported in the dissolved and particulate phases within the Amazon and Yukon Rivers. The percentages transported with particulate species are generally greater than 90 % (data from Gibbs 1977; figure from Miller and Orbock Miller 2007)

Table 1.1 Ratio of dissolved to total elemental transport in rivers

Percentage (%)	Elements	Dominant transport load
1–0.1	Ga, Tm, Lu, Gd, Ti, Er, Nd, Ho, La, Sm, Tb, Yb, Fe, Eu, Ce, Pr, Al	Particulate phase
10–1	P, Ni, Si, Rb, U, Co, Mn, Cr, Th, Pb, V, Cs	Particulate phase
50–10	Li, N[a], Sb, As, Mg, B, Mo, F[a], Cu, Zn, Ba, K	Mixed aqueous and particulate phase
90–50	Br, I[a], S[a], Cl[a], Ca, Na, Sr	Aqueous phase

Lower percentages indicate a greater proportion within the particulate phase
[a] Estimates based on elemental contents in shales
Adapted from Martin and Meybeck (1979)

fair to say that the field has grown over the past 20-years into a scientific subdiscipline in and of itself as indicated by the publication of multiple books on the topic (e.g., Morrison and Murphy 2006; Murphy and Morrison 2007; Morrison and Sullivan 2007; Hester and Harrison 2008; Mudge 2009; Bergslien 2012), the creation of two scientific journals devoted entirely or partially to the field (*Environmental Forensics* and *Environmental Science: Processes and Impacts*), and the creation of university degree programs in Environmental Forensics.

Even a cursory examination of the above mentioned books shows that tracer technology has become an integral part of Environmental Forensics. For example, tracers have not only been used to determine sediment/contaminant provenance in riverine systems, but they have now been applied to address a host of other issues ranging from the redistribution of sediment on hillslopes, to the exchange rates and residence times of sediment within the channel, to the rates of sediment movement to the catchment mouth, to the biogeochemical cycling of contaminants within the aquatic environment. Tracers have also been used to provide retrospective information on geomorphic and geochemical processes and process rates over the past several decades to centuries, data that cannot be obtained by traditional monitoring programs. For example, geochemical tracers may be incorporated into channel, floodplain, or terrace deposits where their analysis may be used to unraveled such things as the timing and history of contaminant influx to rivers and/or the dispersal pathways through which contaminants are distributed along the river (Miller 2013; Miller and Orbock Miller 2007). In light of the above, tracers can be used to decipher potential environmental impacts of sediment and sediment-associated contaminants on river systems, and determine potentially responsible parties associated with these impacts.

The primary objectives of the following chapters are to (1) provide an in depth discussion of the theory, methodology, and application of environmental tracer and fingerprinting methods that have and are currently being used to address the source, transport, and deposition of sediment and sediment-associated contaminants within river systems, and (2) provide an analysis of the strengths and limitations of the examined techniques in terms of their temporal and spatial resolution, data requirements, and inherent uncertainties in the generated results. We will focus on the use of natural and anthropogenic geochemical tracers that currently exist within surficial geological materials, rather than 'particle-tracking' techniques. It is important to recognize that our intent is not to replace other forms of analyses of the sediment system, but to show how tracer/fingerprinting studies can be used to gain insights into system functions that would not otherwise be possible. In fact, significant attention is given to ways in which fingerprinting and tracer technologies may be integrated with other hydrological, geochemical, geomorphic, and stratigraphic techniques to address the complexity inherent in the dispersal of sediment and sediment-contaminated materials through riverine environments. Given that the use of tracers to address legislative or legal issues will undoubtedly increase in the coming years, we will, where possible, address a number of topics that are critical to environmental forensics, including whether the methods represent (1) valid and testable approaches that have gained widespread acceptance through the peer review process, (2) generate results with quantifiable errors or levels of uncertainty, and (3) can be easily understood by individuals who may not have a scientific background (e.g., as a judge or jury).

1.2 Book Format and Overview

We begin our discussion of environmental tracers in Chap. 2 with an overview of what is typically referred to as geochemical fingerprinting. The fingerprinting approach is typically focused on sediment, rather than contaminants (although contaminants attached to those sediments may also be of interest). Moreover, fingerprinting studies are typically aimed at deciphering the relative contributions of sediment from a set of sources, which may be defined spatially (e.g., by the underlying geology or land-use/land-cover type), or according to the process that delivers sediment to the channel (e.g., sheet, rill, gully, or bank erosion). Although a number of novel methods have been put forth (e.g., Poulenard et al. 2009, 2012), source determinations are primarily conducted by fingerprinting sediment sources on the basis of the physical and/or chemical characteristics of the surface sediments and then comparing selected types of river sediment to the fingerprint. Chapter 2 begins with an overview of the assumptions inherent in geochemical fingerprinting, before turning to a more detailed discussion of the uncertainties involved in the methods. More specifically, it examines the various approaches that have recently been developed to quantify and reduce uncertainties in the utilized approach. We then leave our discussion of geochemical tracers as defined by elemental concentrations and turn our attention in Chap. 3 to short-lived fallout radionuclides (FRNs) (e.g., ^{210}Pb, ^{137}Cs, ^{7}Be). These isotopic tracers are somewhat unique in that they cannot only be used to assess sediment provenance (often by source type, e.g., sheet, rill, gully, and bank erosion), but can be used to determine sediment transfer rates within and between specific landscape units, sediment exchange and residence times within the channel bed, and the age of alluvial deposits, among a host of other processes. FRNs, then, may be applied at much smaller spatial scales.

In Chap. 4, we turn our attention to radiogenic isotopes, another group of tracers that have been extensively used to source sediment and sediment-associated contaminants in river systems. Some radiogenic isotopes (e.g., Sr and Nd) have been extensively applied to source sediment, whereas others (e.g., Pb) have been used less as a direct tracer of sediment, but rather are primarily used to track Pb contaminated materials. The spatial scale of application varies with the specific isotopes, but as a group ranges from an individual reach to the global tracking of contaminated dust and aerosols. Interestingly, there is little overlap in the literature on the use of radiogenic isotopes to trace sediments versus contaminated materials, nor are the radiogenic isotopes extensively used as a fingerprint in the analysis of non-point source sediment provenance. While we focus on the past use of selected radiogenic isotopes as tracers, the hope is that the text will provide insights into their wider application to sediment-associated riverine problems.

In the final chapter (Chap. 5), we examine the potential use of a set of stable isotopes that we have only recently been able to analyze in Earth and biological materials at the levels required for their use as environmental tracers. Study of these 'non-traditional' isotopes is therefore limited, but increasing at a nearly exponential rate as they may be effective at determining both the provenance and biogeochemical cycling of trace metals/metalloids. It would, in the available space, be impossible to cover all of these "non-traditional" isotopes here. Thus, we focus on four (including Cd, Cu, Hg, and Zn) that appear to be particularly applicable to rivers and alluvial sediments.

An appendix containing commonly used acronyms (abbreviations), unit conversions and elemental data also is included for convenience.

References

Bergslien E (2012) An introduction to forensic geosciences. Wiley-Blackwell, Chichester

Black KS, Athey S, Wilson P, Evans D (2007) The use of particle tracking in sediment transport studies. In: Balson PS, Collins MB (eds) Coastal and shelf sediment transport, vol 274. Geological Society of London Special Publication, London, pp 73–91

Boswell P (1933) On the mineralogy of sedimentary rocks: a series of essays and a bibliography. Thomas Murphy & Co, London

Gibbs R (1977) Transport phases of transition metals in the Amazon and Yukon rivers. Geol Soc Am Bull 88:829–843

Guzmán G, Barrn V, Gómez JA (2010) Evaluation of magnetic iron oxides as sediment tracers in water erosion experiments. Catena 82:126–133

Haddad R (2004) Invited editorial: what is environmental forensics? Environ Forensics 5:3

Hester R, Harrison R (2008) Environmental forensics: issues in environmental science and technology, vol 26. Royal Society of Chemistry RSC Publishing, Cambridge

Horowitz A (1991) A primer on sediment-trace element chemistry, 2nd edn. Lewis, Chelsea

Horowitz AJ, Elrick KA (1988) Interpretation of bed sediment trace metal data: methods of dealing with the grain size effect. In: Lichtenberg JJ, Winter JA, Weber CC, Fradkin L (eds) Chemical and biological characterization of sludges, sediments, dredge spoils, and drilling muds. American Society for Testing and Materials, Philadelphia, pp 114–128

Hu GQ, Dong YJ, Wang H, Qiu XK, Wang YH (2011) Laboratory testing of magnetic tracers for soil erosion. Pedosphere 21:328–338

Kimoto A, Nearing MA, Shipitalo MJ, Polyakov VO (2006) Multi-year tracking of sediment sources in a small agricultural watershed using rare earth elements. Earth Surf Proc Land 31:1763–1774

Martin J.-M., Meybeck M (1979) Elemental mass balance of material carried by major world rivers. Mar Chem 7:173–206

Mentler A, Strauss P, Schomakers J, Hann S, Köllensberger G, Ottner F (2009) Organophilic clays as a tracer to determine erosion processes. EGU General Assembly 2009. Geophys Res Abstr 11:EGU2009-13192

Meybeck M, Helmer R (1989) The quality of rivers: from pristine stage to global pollution. Palaeogeogr Palaeocl, Glob Planet Change Sect 75:283–309

Miller JR (2013) The forensic assessment of riverine contamination by historic and modern mining activity in the 21st century. In: Cliff D (ed) Taking health and safety in the mining industry into the 21st century—innovative solutions to difficult problems. Spec Issue Miner 3:192–246

Miller JR, Orbock Miller SM (2007) Contaminated rivers: a geomorphological-geochemical approach to site assessment and remediation. Springer, Berlin

Morrison R, Murphy B (2006) Environmental forensics: contaminant specific guide. Elsevier Academic Press, Burlington

Morrison RD, O'Sullivan G (2012) Environmental forensics. In: Proceedings of the 2011 INEF conference. Special Publication No 338, The Royal Society of Chemistry, RSC Publishing, Cambridge, p 345

Mudge S (2009) Methods in environmental forensics. CRC Press, Taylor & Francis Group, Baca Raton

Mukundan R, Walling D, Gellis A, Slattery M, Radcliffe D (2012) Sediment source fingerprinting: transforming from a research tool to a management tool. JAWRA 48:1241–1257

Murphy B, Morrison R (2007) Introduction to environmental forensics, 2nd edn. Elsevier Academic Press, Burlington

Osterkamp WR (2004) An invitation to participate in a North American sediment-monitoring network. Eos Trans AGU 85:386

Parsons AJ, Wainwright J, Abrahams AD (1993) Tracing sediment movement in interrill overland flow on a semi-arid grassland hillslope using magnetic susceptibility. Earth Surf Process Landf 18:721–732

Pimentel D, Harvey C, Resosudarmo P, Sinclair K, Kurz D, McNair M, Crist S, Shpritz L, Fitton L, Saffouri R, Blair R (1995) Environmental and economic costs of soil erosion and conservation benefits. Science 267:1117–1123

Poulenard J, Perrette Y, Fanget F, Quetin P, Trevisan D, Dorioz JM (2009) Infrared spectroscopy tracing of sediment sources in a small rural watershed, French Alps. Sci Total Environ 407:2808–2819

Poulenard J, Legout C, Nmery J, Bramorski J, Navratil O, Douchin A, Fanget B, Perrette Y, Evrard O, Esteves M (2012) Tracing sediment sources during floods using diffuse reflectance infrared fourier transform spectrometry (DRIFTS): a case study in a highly erosive mountainous catchment, Southern French Alps. J Hydrol 414415:452–462

Schumm S (1977) The fluvial system. Wiley, New York

Spencer KL, Droppo IG, He C, Grapentine L, Exall K (2011) A novel tracer technique for the assessment of fine sediment dynamics in urban water management systems. Water Res 45:2595–2606

USEPA (2013) Watershed assessment, tracking, and environmental results: national summary of state information. http://ofmpub.epa.gov/waters10/attains_index.control, accessed 7 Oct 2013

Walling DE, Golosov V, Olley J (2013) Introduction to the special issue: tracer applications in sediment research. Hydrol Process 27:775–780

Weltje GJ (2012) Quantitative models of sediment generation and provenance: state of the art and future developments. Sediment Geol 280:4–20

Wenning R, Simmons K (2000) Editorial. Environ Forensics 1:1

Zhang XC, Nearing MA, Polyakov VO, Friedrich JM (2003) Using rare-earth oxide tracers for studying soil erosion dynamics. Soil Sci Soc Am J 67:279–288

Chapter 2
Geochemical Fingerprinting

Abstract Use of geochemical fingerprinting methods to determine sediment provenance has progressively increased since the late 1990s, and is now considered by many investigators as the method of choice to quantify sediment source contributions at the catchment scale. Application of geochemical fingerprinting largely rests on four factors: (1) the inability of other techniques (e.g., sediment load monitoring, photogrammetric methods, and mathematical modeling approaches) to effectively determine sediment provenance at the required spatial scales, (2) improvements in analytical methods that allow for the analysis of large numbers of samples for a wide range of elements, (3) the modification of the utilized statistical methods (e.g., inverse/unmixing models) to more effectively account for uncertainty in the modeled results, and (4) the ability to apply the methods to historic sedimentary deposits retrospectively to determine changes in sediment provenance at a site through time. In this chapter, we focus on the application of geochemical fingerprinting to contemporary river sediments as well as alluvial deposits that are less than about 150 years old. Our intent is not simply to summarize the voluminous and growing body of literature on the subject, but to document the strengths, weaknesses, and uncertainty inherent in the approach.

Keywords Geochemical fingerprinting · Sediment provenance · Unmixing models · Model uncertainty

2.1 Introduction

In order to mitigate the impacts of sediment and sediment-associated contaminants on aquatic ecosystems, one must first determine from where the sediment is derived. Once identified, the predominate sediment sources can be targeted using the often limited financial resources available. While conceptually simple, identifying sediment sources is not as easy as you might think. For example, the use of site specific monitoring of sediment loads to determine the source of sediments to a water body has proven to be a costly, labor intensive, long-term process with a spatial resolution limited by the number of monitoring sites that can be effectively maintained for significant periods of time. An alternative approach is to identify upland areas that are being eroded and then quantify the rate at which sediment is being removed. Such

© The Author(s) 2015
J.R. Miller et al., *Application of Geochemical Tracers to Fluvial Sediment*,
SpringerBriefs in Earth Sciences, DOI 10.1007/978-3-319-13221-1_2

methods have been aided in recent years by technological advances in surveying, remote sensing, and photogrammetric techniques that have improved our ability to document temporal and spatial patterns in erosion (Collins and Walling 2004). Collins and Walling (2004) point out, however, that these methods fail to determine the degree to which sediment sources are connected to the river and the inherent uncertainty in routing sediment from the source to the channel. To overcome the problems inherent in the direct measurement of sediment loads or upland erosion rates, distributed modeling routines have been used, but these complex algorithms require the collection and compilation of significant input and validation data, and often have difficulties apportioning riverine sediments to individual sources (Collins and Walling 2004). As a result, investigators have turned in recent years to the use of physical and geochemical tracers, which can be applied relatively rapidly to gain insights into the source of sediment and sediment-associated contaminants within a catchment.

The specifics of the fingerprinting approach vary widely, as do the parameters that have been used as tracers to determine the source of sediments contained within a river or its associated features (e.g., floodplain, reservoir, riparian wetland, etc.) (for a review, see D'Haen et al. 2012). Table 2.1, while far from exhaustive, shows the most commonly utilized parameters with regards to riverine systems. The applicability of these methods varies according to (1) the grain size fractions to which they can be applied (i.e., gravel, sand, or silt and clay-sized material), and (2) the temporal and spatial scale for which they can be used (D'Haen et al. 2012). To date, an overwhelming majority of source ascription studies at the catchment scale have focused on fine-grained sediments ($< \sim 63 \, \mu m$ in size) eroded from diffuse upland areas in response to either natural or anthropogenic disturbances (e.g., wildfires, deforestation or timber harvests, agricultural practices, and urban/exurban development). The focus on fine sediment, as noted in Chap. 1, reflects both its direct impacts on riverine ecosystems (Wood and Armitage 1997; Armstrong et al. 2003; Syvitski et al. 2005; Bo et al. 2007; Kemp et al. 2011) and its chemically reactive nature, which allows for a wide range of contaminants (e.g., nutrients, agricultural chemicals, and trace metals and metalloids) to be carried from upland areas to the drainage network (Horowitz 1991; Collins et al. 2005; Miller and Orbock Miller 2007). The movement of nutrients from agricultural lands to rivers, reservoirs, and lakes, for example, is often a significant issue in rural areas, and can lead to severe cases of eutrophication (Fig. 2.1). With regards to fine-grained sediments, geochemical tracers, fallout radionuclides (FRNs), and mineral magnetic properties have been most extensively utilized in provenance studies of both historical (50–10,000 ybp) and contemporary (<50 ybp) sediments (D'Haen et al. 2012) (Fig. 2.2).

In this chapter, we focus on a specific methodological approach often referred to as geochemical fingerprinting to determine the provenance of sediments suspended within the water column or contained within alluvial deposits that are less than about 150 years old. The catchment-scale approach involves two primary components: (1) the identification of a set of sediment-associated geochemical parameters (i.e., a fingerprint) that can be used to discriminate between the sediments of variously defined sediment sources, and (2) the estimation of the relative proportion of sediment from each of the individual sources that comprise suspended sediments (or other type

Table 2.1 Tracer types and representative studies that have utilized them (adapted from D'Haen et al. 2012 and Guzmán et al. 2013)

Tracer	Representative references
Physical tracers	
Particle color	Grimshaw and Lewin (1980), Giosan et al. (2002), Krein et al. (2003), Croft and Pye (2004), Martínez-Carreras et al. (2010)
Grain size distribution	Dudley and Smalldon (1978), Kurashige and Fusejima (1997), Stuut et al. (2002), Weltje and Prins (2003), Weltje and Prins (2007), Weltje (2012)
Grain morphology and texture	de Boer and Crosby (1995), de Boer et al. (2000), Cardona et al. (2005), Madhavaraju et al. (2009)
Magnetic properties (χ_{Lf}, χ_{Hf}, χ_{Fdep}, ARM, IRM, HIRM)[a]	Yu and Oldfield (1993), Caitcheon (1998), Oldfield et al. (1999), Slattery et al. (2000), Dearing et al. (2001), Jenkins et al. (2002), Morton and Hallsworth (1994), Oldfield (2007), Zhang et al. (2008), Maher et al. (2009), Hatfield and Maher (2009), Armstrong et al. (2010), Guzmán et al. (2010)
Mineralogical tracers	
Mineralogy	Abu-Zeid et al. (2001), Arribas et al. (2000), Pirrie et al. (2004), Pye (2004), Benedetti et al. (2006)
Heavy minerals	Basu and Molinaroli (1991), Damiani and Giorgetti (2008), Oszczypko and Salata (2005), Vologina et al. (2007), Hardy et al. (2010)
Clay minerals	Eberl (2004), Gingele and De Deckker (2005)
Cathodo-luminescence quartz	Gotze et al. (2001), Bernet and Bassett (2005), Gotte and Richter (2006)
Geochemical and biogeochemical tracers	
Major elements	Rollinson (1993), Douglas et al. (2003)
Rare earth elements	Mahler et al. (1998), Zhang et al. (2008), Polyakov and Nearing (2004), Polyakov et al. (2009), Kimoto et al. (2006), Lee et al. (2008), Deasy and Quinton (2010), Yang et al. (2008), Wude et al. (2008), Singh (2009), Xu et al. (2009), Collins et al. (2013), Miller et al. (2013)
Trace metals metalloids (Cd, Cu, Pb, Zn, As)	Collins et al. (1997a), Collins et al. (1998), Collins et al. (2010a), Collins et al. (2012), Collins et al. (2013), Miller et al. (2005), Hallsworth and Chisholm (2008), Decou et al. (2009), Grimes et al. (2007), Rowan et al. (2012), Massoudieh et al. (2013), Zhang et al. (2012)
Elemental ratios (e.g., Cu/Pb; Si/Al; Pb/Al)	Wang et al. (2009), Rowan et al. (2012)
Fallout radionuclides (^{137}Cs, ^{210}Pb, ^{7}Be, ^{239,240}Pu)	Wallbrink and Murray (1993), Walling and He (1999), Walling et al. (1999), Walling et al. (2009), Wallbrink et al. (2002), Nagle et al. (2007), Mabit et al. (2008), Ritchie and Ritchie (2008), Wilkinson et al. (2009), Evrard et al. (2010), Parsons and Foster (2011), Taylor et al. (2012), Gaspar et al. (2013), Golosov et al. (2013), Walling (2013), Wilkinson et al. (2013)
Isotopic ratios (δ^{13}N, δ^{13}C, δ^{87}Sr, ^{204}Pb/^{206}Pb, etc.)	Douglas et al. (1995), Douglas et al. (2003), Gingele and De Deckker (2005), Lee et al. (2008), Yang et al. (2007), Fox and Papanicolaou (2008a), Fox and Papanicolaou (2008b), Alt-Epping et al. (2009), Mukundan et al. (2010)
Biogeochemical (N, C, P)	Hasholt (1988), Hillier (2001), Fox and Papanicolaou (2008b), Alt-Epping et al. (2009), Hancock and Revill (2013)
Mineral ages (zircon, monazite, muscovite)	Gleason et al. (2007), Kirkland et al. (2009), Veevers and Saeed (2007), Reynolds et al. (2009), Amidon et al. (2005), Morton et al. (2008)

[a] χ Magnetic susceptibility (low, high frequency and frequency dependent); *ARM* Anhysteretic remanent magnetization; *IRM* isothermal remanent magnetization; *HIRM* derived remanence parameters

Fig. 2.1 Eutrophication in a reservoir within the KwaZulu-Natal region of South Africa

of alluvial material) within the river. The latter is accomplished by comparing the geochemical parameters that make up the fingerprint in the source sediments to that of the riverine material. The use of such geochemical fingerprinting techniques has increased dramatically since the late 1990s. In fact, many investigators now consider geochemical fingerprinting the method of choice with respect to diffuse sediment sources. The increased use of geochemical fingerprinting is due, in part, to recent advances in analytical instrumentation that allow for large numbers of elements to be analyzed in a large number of samples in a relatively short period of time. These analytical advances have been accompanied by the enhancement of source ascription methods that provide for a more detailed and quantitative understanding of the uncertainty inherent in the derived results. The intent of our analysis herein is not simply to summarize the voluminous and growing body of literature on the subject, but to document the strengths, weaknesses, and uncertainty inherent in the approach in general, and specific methods in particular.

2.2 Conceptual Model and Inherent Assumptions

Upstream portions of the riverine sediment-dispersal system are characterized by a network of channels and their associated hillslopes, both of which serve as zones of sediment production (Figs. 1.2 and 2.3). Hillslope areas can be geographically

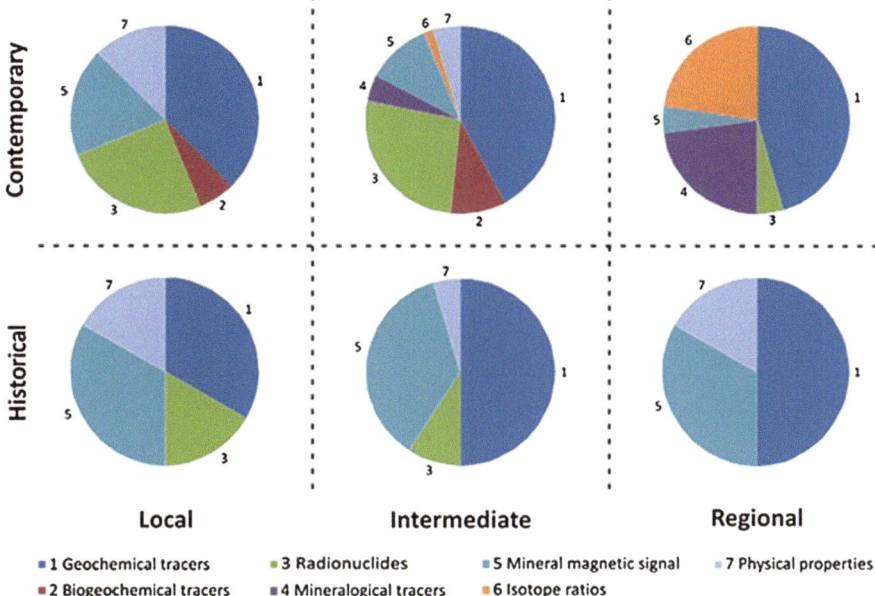

Fig. 2.2 Relative use of tracer types identified and categorized on the basis of temporal and spatial scale by D'Haen et al. (2012) for determining the provenance of fine-grained sediment in alluvial deposits. Spatial scales defined as local ($<10\,km^2$), intermediate (10–$10,000\,km^2$) and Regional ($>10,000\,km^2$); temporal scales subdivided into contemporary ($<50\,ybp$), and historical (50–$10,000\,ypb$) (from D'Haen et al. 2012). Examples of tracers associated with each category of tracer provided in Table 2.1

subdivided into units on the basis of the underlying geology, soil type, etc., each unit defining a distinct sediment source. Sediment sources can also be defined according the processes of sediment generation (e.g., whether the sediment was derived from sheet, rill, gully or bank erosion). Source areas defined according to the generating process are often referred to as source types. Particles eroded from these defined source areas or types are transported, often intermittently, through a channel/valley network to a downstream depositional basin that serves as a long-term sediment repository (Weltje 2012). In the process, particles from all of the source areas are combined such that the sediments within the channel represent a mixture of particles from all of the source areas in the basin. The physical and geochemical composition of this sediment mixture (which we will refer to as river sediment) is a function of the composition of the source area sediments and the relative amount of sediment that each source area contributes; if both are known, then it is possible to predict the composition of the mixture. Mathematically, this predictive calculation is considered a linear forward problem (Weltje 2012). More commonly, however, the objective is to determine the relative volume of material supplied to a particular type of river sediment (suspended load, channel bed material, floodplain deposit, etc.) from each sediment source. This type of calculation represents a linear inverse

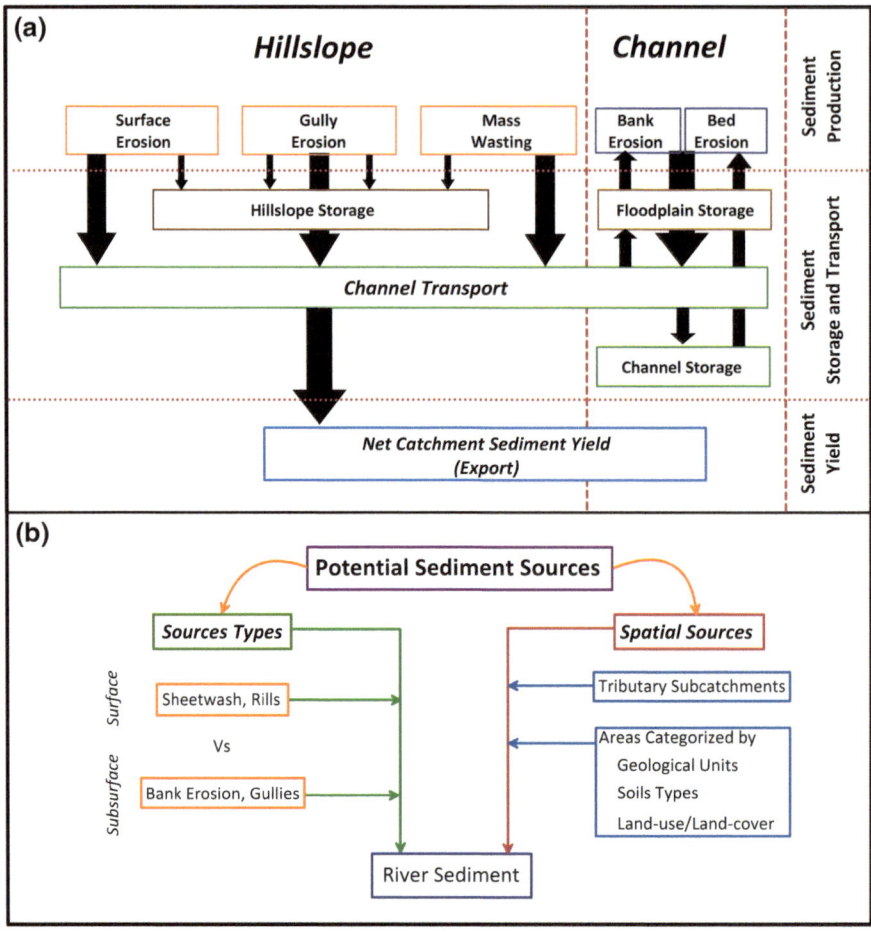

Fig. 2.3 a Flow diagram showing typical sources and pathways of sediment movement within upland areas (modified from USEPA 1999). **b** Classification of the potential sources of riverine sediments commonly defined for geochemical fingerprinting studies

problem when the number of sources and their characteristics are known (Weltje 2012); it is typically solved using a statistically based inverse or unmixing model. In essence, inverse modeling requires that the composition of the source materials be known for a selected set of physical and/or chemical properties, and then defines the mixing proportions from each source that best fits the observed composition of the studied river sediment (Weltje 2012).

The application of inverse modeling to riverine sediments is complicated by the fact that the composition of the alluvial sediment reflects a wide range of physical and geochemical processes in addition to the simple mixing of particles from the source areas (Johnnson 1993; Weltje and von Eynatten 2004). Of particular significance are hydraulic sorting processes in which the original population of particles from a source area is modified by the selective entrainment (erosion), transport, and deposition of grains as they are dispersed through the system according to their size, density, and shape (Knighton 1998; Miller and Orbock Miller 2007). Mechanical and chemical weathering processes also lead to modifications in the initial grain population. The net effect of these processes is that sediment of different size, shape, and density within the river is transported downstream at different rates, often producing a downstream fining in particle size. Sediment also may be transported by different methods (e.g., by suspended and bedload process) (Weltje 2012), or be partitioned by the flow into distinct depositional units at a given site (e.g., pools, riffles, point bars, floodplains, etc.) (Miller and Orbock Miller 2007). The river and source area sediments, then, may represent two very different populations of particles. Suspended sediments, for example, may only represent the finest materials within the source areas, and their mineralogy would be expected to differ from that of the bulk material. Moreover, the geochemical properties of the river and source area sediments are likely to differ as fine-grained particles characterized by large surface areas and high surface charge tend to be more reactive and have a greater potential to collect, concentrate, and retain ions (e.g., trace metals).

Modification of the source area sediments during dispersal by physical and chemical processes is important because an assumption inherent in inverse modeling is that the physical and geochemical composition of river sediment differs from a specific source area only because it has been mixed with sediment from another source area(s). Thus, physical and chemical modifications of the source area sediments during erosion, transport and/or deposition must be eliminated, or at least limited, to effectively use inverse modeling. Provenance studies, particularly those aimed at determining the provenance of sediments in stratified rocks, often deal with these modifications using a concept that Weltje (2004) referred to as transport invariance. The concept assumes that particles with similar sizes, shapes, and densities will be entrained, transported and deposited under similar conditions. Thus, by comparing particles from the source areas and the river that fall within a narrowly defined range of sediment size, density or shape, the potential, transport-related modifications can be reduced, and the composition of the river sediment will primarily reflect the relative mixing of sediment from the various source areas. As we will see below, approaches other than the analysis of transport invariant populations have also been proposed to account for the physical and geochemical modification of the source sediments. The point to be made here is that a determination of the provenance of the bulk sediment (consisting of a wide range of particle sizes) may require the combined analysis of multiple size ranges. In fact, it is quite possible that the predominant source(s) of sediment found within the river may vary as a function of particle size (Miller et al. 2013). Sandstone strata, or the soils developed within it, for example, are likely to contribute more sand-sized sediment to a river than a shale and its associated soils. Determining the provenance of multiple size fractions can be time consuming and expensive. Thus, most studies of sediment provenance focus on the particle size fraction that is of importance to the question at hand. For the majority of the geochemical fingerprinting studies, the focus has been on relatively fine-grained sediment ($<63\,\mu m$) as it is this size fraction that forms a significant portion of the suspended load, is largely responsible for decreasing water quality, and is chemically reactive, thereby serving as an important conveyor of hydrophobic contaminants. Sand-sized sediment, however, may also be of importance. Within many of the gravel-bed rivers of the Southern Appalachian Mountains of the southeastern U.S., for instance, aquatic habitats are predominantly affected by the deposition of sand-sized sediments on the channel bed, and their infiltration into the interstitial spaces between gravel sized clasts.

Another fundamental problem inherent in inverse mixing models is the potential for sediments to be eroded from a defined sediment source, transported downvalley and temporarily deposited within the channel (or some other unit) before being 'remobilized'. These reworked sediments are often difficult to recognize (Weltje 2012); thus, it is generally assumed that the source area sediment travels directly from its point of detachment to its point of sampling. The degree to which this assumption is violated depends largely on the size of the basin and the degree of physical connectivity that exists along the drainage network; the chances of determining the ultimate source of sediment, and not its proximal one, decreases with increasing catchment size and decreasing connectivity (Miller et al. 2013).

Inverse modeling, as defined above, is aimed at determining the relative contribution of sediment from defined source areas to a specific type of river sediment. Emphasis is placed on the composition of the river sediment and the origin of the particles contained within it. Some geochemical fingerprinting studies, however, propose a slightly different objective: to assess the relative amount of sediment eroded from the defined source areas or source types. The difference between these two objectives is subtle, but important. When the goal is to determine the amount of sediment eroded from each of the sediment sources, an additional assumption is applied to geochemical fingerprinting. It must be assumed that the sediment leaves all sources at the same time and is transported downstream at an equal rate so that it arrives at the sampling point simultaneously. This assumption is often violated by differences in the proximity of a source to the sampled depositional area, or by differences in the rate at which particles of differing size or shape are transported downstream (the transport variance problem). Take, for example, a 2 cm thick sample collected from the surface of a floodplain that received sediment from two upstream sources. One source is located immediately adjacent to the sampling site, whereas the other is located a considerable distance upstream. Also assume that equal amounts of material are eroded from both sediment sources, and the rate of sediment deposition from both sources is the same. At the onset of the runoff event sediment from the closest source will reach the site first; thus, the lower portions of the 2 cm thick sampling interval will be composed of material from only this source. As the event continues, material from the other source reaches the site, and equal proportions of sediment from both sources will be deposited at the site. If the entire 2 cm of sediment is not composed of a single event, the other events will follow the same pattern until 2 cm of sediment has been accumulated. When the inverse/mixing model is applied to the sample, it will correctly indicate that a larger relative percent of sediment was derived from the closest site over the timeframe represented by the 2 cm increment (this is the objective of the inverse modeling as defined earlier). Thus, sediment provenance with respect to the deposit has been correctly assessed within the errors inherent in the statistical analysis. However, if the intent is determine the relative amount of sediment eroded from the two source areas, the results will be biased such that the model will overestimate the amount of material eroded from the closest source. Differences in particle transport rates produced by varying particle sizes can lead to similarly biased results.

In the case where elemental concentrations are used as geochemical fingerprints, there is also an assumption that the elements exhibit conservative behavior. That is, the elements selected as a fingerprint move with the sediment and are not lost from the system. This follows because inverse/mixing models represent a form of mass balance analysis. Thus, elements that tend to be mobile within aquatic systems and possess lower affinities for particulate matter generally serve as poor fingerprints.

2.3 Methodological Approach

While the specific methods used in geochemical fingerprinting varies from one investigator to the next, the general approach involves the completion of five key steps (after Zhang et al. 2012): (1) delineation and characterization of sediment sources within the catchment, (2) determination of the fingerprinting properties that most effectively identify and discriminate between sediments of the defined sources, (3) collection and characterization of river sediment, selected on the basis of the time-frame under consideration, (4) determination of sediment provenance using numerical modeling procedures, and (5) assessment of the uncertainty inherent in the modeling results. These steps are discussed in detail below.

2.3.1 Source Delineation

The first step in any fingerprinting analysis is to define the primary sediment sources within the catchment that may be of interest. Historically, sediment sources have been subdivided into two main categories: upland (hillslope) sediments, and channel bed and banks sediments (Fig. 2.3a). Both types of sediment may be eroded and transported to the water body by one or more geomorphic processes.

For fingerprinting analyses, upland sources are often subdivided further on the basis of the spatial extent and location of mapped geological units (Collins et al. 1997a; Walling et al. 1999; Douglas et al. 2003; Miller et al. 2005), soil types (Miller et al. 2013), land-use/land-cover categories (Collins 1995; Walling and Woodward 1995; Russell et al. 2001; Miller et al. 2013), or contributing tributary areas (Klages and Hsieh 1975; Collins et al. 1997b, 2009, 2010a) (Fig. 2.3b).

This spatially defined source approach is plagued by several problems. First, soil erosion is not only a function of soil type, land-use, or the underlying geology, but varies as a function of factors such as topography and process. Agricultural pastures, for example, may be eroded in steep upland areas by sheet and rill processes and on low-relief floodplains adjacent to the channel by advancing headcuts associated with gullies (Fox and Papanicolaou 2008b). Thus, spatially defined sources fail to directly identify the geomorphic processes responsible for sediment generation. Second, soil types and land-use/land-cover categories are often transitional to one another, confounding their spatial delineation within the catchment as well as the geochemical differences in their sediments (Rowan et al. 2012). Differences in the underlying geological units may also complicate the issue. Third, recent changes from one land-use/land cover type to another may limit the ability of geochemical parameters to distinguish between sediment source areas (Miller et al. 2013). In other words, the geochemistry of the sediment sources may reflect both its current and past land cover history, making it difficult to distinguish between sediments associated with the various land-use/land-cover categories.

In light of the above, an alternative method of defining sediment sources is by the erosional process through which the sediments are generated and delivered to the river. Referred to as the 'source type', a distinction is most often made between sediments generated near or at the ground surface in upland (hillslope) areas by sheet or rill erosion and sediment derived from the 'subsurface' by means of gully or bank erosion (Fig. 2.3) (Walling and Peart 1979; Gellis et al. 2009; Gellis and Walling 2011; Massoudieh et al. 2013). Differentiation between surface and subsurface sediments requires the use of geochemical parameters that differ as a function of depth below the ground surface, such as organic matter or short-lived radionuclides (described in the next chapter).

As neither the spatial or process approach to defining sediment sources is ideal on its own, it is not uncommon for investigators to combined the two methods, thereby defining sediment sources on the basis of both spatial and type categories (Russell et al. 2001; Juracek and Ziegler 2009; Wilkinson et al. 2009), particularly for catchments less than about 200 km^2 (Mukundan et al. 2012). Within larger catchments, the heterogeneity of sediment source properties defined by land-use, soil type, or geomorphic process is likely to increase, making it more difficult to distinguish between the sources and hindering source ascription (Collins et al. 1998). In addition, sediment contributions from relatively minor sources, which may still cover large areas, could be underestimated (Mukundan et al. 2012). As a result, the application of geochemical fingerprinting methods to large basins (>500 km^2) is more difficult, although a number of studies have shown that sediment sources may be effectively defined according to the underlying geological units within the catchment (Walling et al. 1999; Bottrill et al. 2000; Douglas et al. 2003) or by tributary catchment areas (Collins et al. 1996; Walling et al. 1999), both of which tend to exhibit less property heterogeneity than sediments defined according to land-use, soil type, or erosion process (Collins et al. 2012; Mukundan et al. 2012; Wilkinson et al. 2013).

2.3.2 Collection and Characterization of River Sediment

A wide range of river sediments have been targeted for geochemical fingerprinting (Fig. 2.4). The sediments which are selected dictate the timeframe under

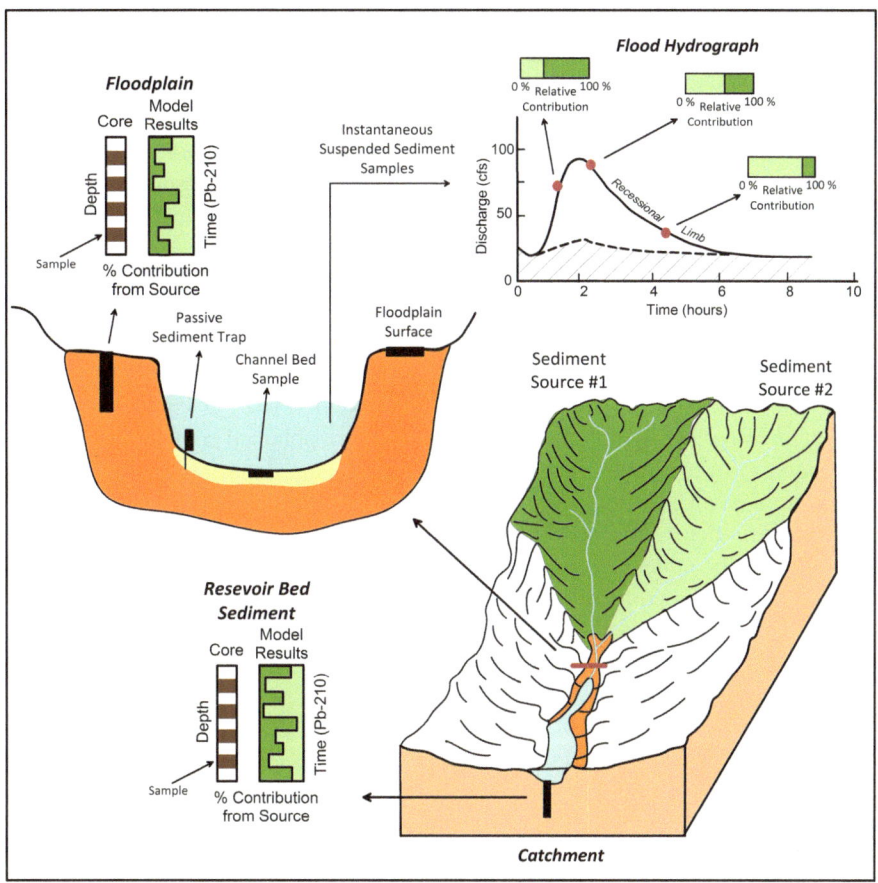

Fig. 2.4 Illustration of the types of river sediments that are collected for the geochemical finger-printing of contemporary and historical sediments. Each type of sample is associated with a specific timeframe of analysis, ranging from instantaneous samples collected during a specific part of a flood hydrograph (*red, filled circles*) to passive and channel bed sediments (periods of months to a few years, *horizontal rectangles*) to floodplain and reservoir sediments (representing periods of years or decades, *vertical rectangles*)

consideration, and may range from recent, instantaneous suspended sediment samples collected during a specific portion of a flood hydrograph to river sediments deposited within floodplains or riparian wetlands during an entire flood and that may be hundreds or even thousands of years old.

Historically, suspended sediments were sampled to assess the contributions of contemporary sediments to the channel from key sources during flood events (e.g., Collins et al. 1997a, 1998, 2001; Peart and Walling 1986; Walling and Woodward 1992, 1995). There has been a growing realization, however, that the fingerprinting of instantaneously collected samples may not be the most time and cost effective method of determining sediment provenance for an entire flood. At issue is the fact that sediment loads typically vary throughout the event, with larger loads tending to be associated with higher flows. The relationships between sediment load and discharge is not always perfect, however, as larger sediment loads may be associated with the rising, peak, or falling stages of a hydrograph. For example, the episodic erosion of easily eroded sediment during the onset of a runoff event often leads to larger loads during the rising stage of the flood, in comparison to the same discharge conditions during the falling stage, producing a phenomena referred to as the first flush (Miller and Orbock Miller 2007). This 'first flush' phenomenon not only demonstrates that the rates of soil erosion vary through the flood, but that erosion varies from one sediment source to another over the landscape at any one time. Thus, the source contributions determined for an instantaneously collected sample will apply only to the portion of the flood that was sampled, rather than for the entire storm (Collins et al. 2001; Massoudieh et al. 2013). To address this issue it is now common to collect an integrated sample in which sediments are obtained continuously or semi-continuously over a longer time span, such as the entire flood. The sediments within these samples can be expected to reflect the averaged contribution from each sediment source within the watershed (Fox and Papanicolaou 2008b). While such integrated samples may be collected using automated, pump-type sampling devices, the need for relatively large sediment sample sizes for geochemical analysis has led to the use of passive samplers or sediment traps (e.g., Phillips et al. 2000; Russell et al. 2001) that collect materials representing the entire storm hydrograph (Massoudieh et al. 2013).

An alternative to the use of these time-integrated sediment traps is to sample the channel bed material (Evrard et al. 2013; Collins et al. 2013) as recent studies have shown that such bed sediments serve as an effective surrogate of continuously collected material over multiple flood events (Miller and Orbock Miller 2007; Horowitz et al. 2012; Collins et al. 2013). Two additional advantages of sampling the channel bed sediment is that it is not necessary to wait for a flood event to conduct the sampling, nor does one have to sample over an extended period of time (Mukundan et al. 2012). The sediment stored in the channel bed may change, however, over time and at an unknown rate. Thus, bed sediment may need to be sampled on more than a single occasion to assess the relative contributions from key sources over, say, an entire year (Collins et al. 2013).

Some investigators have sampled the surface of floodplain deposits (e.g., Collins et al. 2010a, b, 2012). This particular sampling scheme does not assess the sediment loads during low to moderate flood events contained within the channel banks, but rather is used to assess sediment provenance during events capable of inundating the floodplain. The assumption inherent in this approach is that these overbank events transport a majority of the sediment within the catchment, a conclusion reached by studies dating back to the 1960s (e.g., Wolman and Miller 1960). Thus, the results provide a reasonable assessment of sediment source contributions within the catchment by flows that transport, on average, the most sediment (Collins et al. 2012).

While most early studies were aimed at documenting contemporary sediment sources, Mukundan et al. (2012) point out that the same basic approach has been applied to floodplain, reservoir, wetland, and lake deposits to determine the changes in sediment source to a river through time (Fig. 2.4) (Foster et al. 1998; Owens et al. 1999; Walling et al. 2003a, b; Miller et al. 2005, 2013; Pittam et al. 2009; Collins et al. 2010c). Essentially, it is assumed that the sampled deposits represent an historical record of sediment transport within the basin, where the age of the sediment varies as a function of depth below the ground surface. Thus, fingerprinting can be carried out on samples collected at differing depths to reconstruct the changes in sediment provenance to the depositional site through time. The method is useful in that it allows an understanding of the contemporary sediment sources to be placed into an historical framework. It also illustrates that fingerprinting can be used to retrospectively determine the primary sources of sediment to the channel, something that cannot be done using monitoring data.

2.3.3 Identifying Effective Geochemical Fingerprints

Studies of sediment provenance in the 1980s and 1990s often relied on a single para-meter, many of which were based on the physical characteristics of the sediment, such as it grain size distribution, mineralogy, or magnetic properties (Table 2.1). Later investigations, beginning in the late 1990s, showed that erroneous sediment-source area associations were common when only a single fingerprinting parameter was utilized (Collins and Walling 2002). Thus, there was a move to use multiple parameters to fingerprint source area sediments (Collins et al. 1997a, b; Miller et al. 2005; Mukundan et al. 2012; Collins et al. 2010a, 2013; Miller et al. 2013). This compos-ite fingerprinting approach was aided by (1) advances in analytical chemistry that greatly expanded the number and rate for which samples that could be analyzed for a large number of constituents (Walling et al. 2013), and (2) the increased use of mul-tivariate statistical methods to manipulate the composite fingerprinting data, thereby allowing for the quantification of the results. Both factors also increased the use of geochemical parameters as fingerprints, particularly the elemental concentrations of trace metals (Lewin and Wolfenden 1978; Macklin 1985; Knox 1987, 1989; Pass-more and Macklin 1994; Miller et al. 2005, 2013), rare earth elements (Morton 1991; Miller et al. 2013), organic substances (Hasholt 1988), fallout radionuclides (Peart and Walling 1986; Walling and Woodward 1992; Wallbrink and Murray 1993), and various radiogenic or stable isotopes (Douglas et al. 1995). In many cases, the uti-lized geochemical constituents are natural, but in others, investigators have made use of anthropogenic pollutants, such as heavy metals, pesticides and fertilizers (Bravo-Espinoza et al. 2009; Takeda et al. 2004). Takeda et al. (2004), for example, found that while phosphate fertilizers contained 10–200 times more U than soils, they con-tained lower Th concentrations than the soils. Thus, the U/Th ratio proved to be an effective fingerprinting tool (Evrard et al. 2013).

In general, the type of tracer used for a given study depends on how the sediment sources are defined. For example, if the intent is to determine the relative contributions of sediment on the basis of source type (e.g., sheet, rill, gully, and bank erosion), then it will be important to consider constituents that are elevated in surface materials eroded by sheet and rill erosion and low in subsurface materials eroded by gully and bank erosion (or vice versa). Agricultural pesticides or fertilizers may be useful in separating agriculturally related soils from other types of land-use/land-cover. The number of parameters to select also depends on the defined sediment sources because, in general, inverse/unmixing models require n number of parameters to discriminate $n + 1$ sediment sources (Mukundan et al. 2012). However, it is not uncommon to utilize a fingerprint containing more parameters than source areas or types, as described below.

The most common approach at the present time for determining an effective fingerprint is to analyze the source and river sediments for a wide range of constituents and then select the fingerprinting parameters using a multi-step, empirically based process (Fig. 2.5). The nature of these statistical methods is important as they heavily influence the reliability of the fingerprinting results (Walling et al. 2013). The current trend is to use a three step process that eliminates parameters that do not meet certain assumptions inherent in the use of inverse/unmixing models, while identifying the parameters that most effectively discriminate between sediment from the defined sediment sources or source types.

The initial step in this three-part process is to eliminate geochemical parameters from further consideration that do not behave conservatively. Conservative behavior is often determined using simple range tests that ensure that the range of parameter values measured within the sampled river sediment(s) fall within the observed range of values measured for the sampled sediment sources (Billheimer 2001; Phillips and Gregg 2003; Fox and Papanicolaou 2008a; Collins et al. 2012; Wilkinson et al. 2013). This requirement often eliminates relatively soluble elements (e.g., Na, Cl, and P), and those primarily associated with organic matter.

Conservative behavior also requires that there be no enrichment or depletion in parameter values as a result of physical or chemical processes that modify the source area sediment during their dispersal (e.g., by hydraulic sorting or grain weathering) (Mukundan et al. 2012). In essence, the question is whether the sedimentological characteristics of the sediment (e.g., grain size, shape, density, mineralogy) within the source areas and the river can be directly compared as is assumed. Three different approaches have been used to address the issue. Perhaps the most commonly used approach is to analyze and focus on a narrowly defined grain size fraction. This approach is essentially analogous to the transport invariant approach often used to assess the provenance of sediments within lithified strata as described earlier (Weltje 2012). It must be remembered, however, that the results of such a fingerprinting analysis apply only to that grain size fraction. Determining the source of the bulk sample (or other size fractions) will require additional analyses, increasing analytical costs and effort. In addition, the analyses do not provide for an understanding of the actual concentrations in the bulk sample which may be required for other types of environmental assessments (e.g., a pollutant's potential impact on biota).

An alternative approach is to mathematically manipulate the geochemical data obtained from the bulk sediment sample using information collected from a separate subsample of the analyzed sediment. The most common form of manipulation

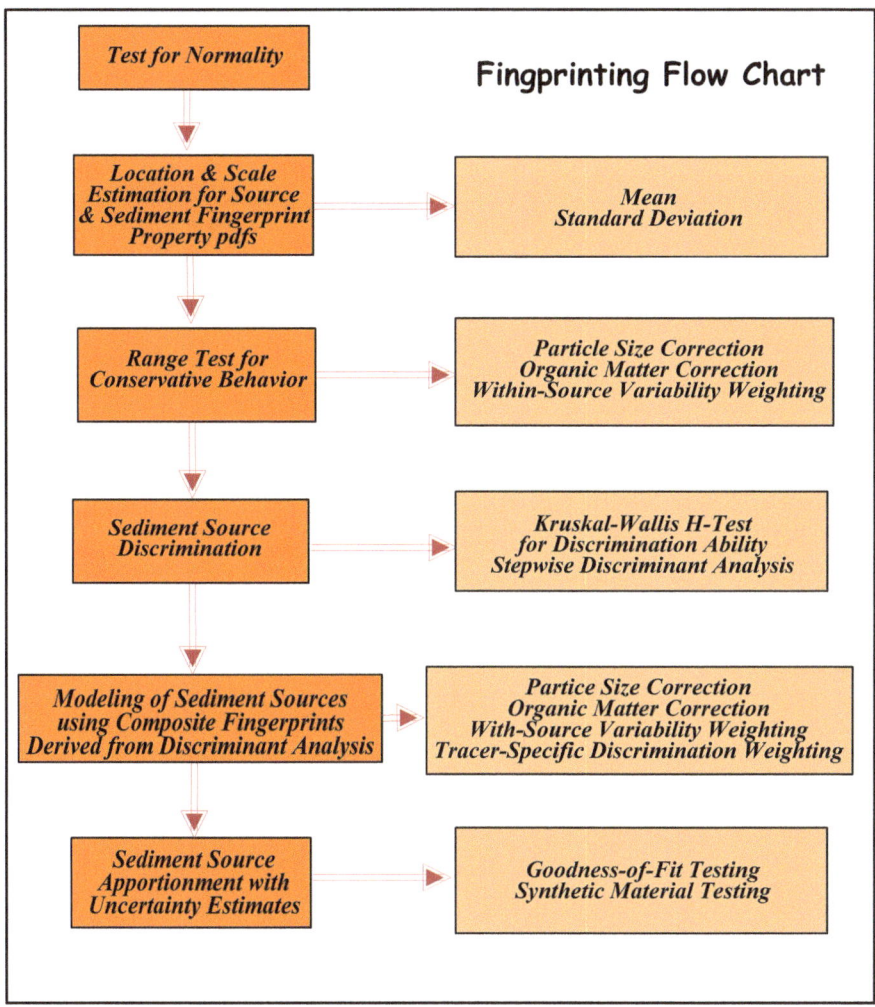

Fig. 2.5 Summary of inverse/mixing model procedure utilized by Collins et al. (2012, 2013) and others

involves the normalization of bulk concentrations (i.e., the concentration measured on the total sediment sample) to account for differences in sediment geochemistry related to particle size and mineralogy. The typical assumption is that certain constituents (e.g., sand-sized sediment or quartz and feldspar grains) act to dilute the concentration of elements associated with the more reactive materials (e.g., fine-grained particles enriched in clay minerals, Fe and Mn oxides and hydroxides, and organic matter). Thus, normalization removes the diluting effects of the non-reactive constituents. Normalized concentration (*NC*) with respect to grain size, for example, estimates the concentration of the sediment if it was entirely composed of fine-grained, chemically reactive material, and is performed using the following equation:

$$NC = DF \cdot BC \tag{2.1}$$

where BC is the bulk concentration and the dilution factor, DF, is calculated as:

$$DF = \frac{100}{RS} \tag{2.2}$$

where RS is the percentage of reactive sediment of a given size range.

A significant disadvantage of the approach is that the normalized data do not necessarily reflect the actual chemical concentrations within the sampled sediments for the selected size range, particularly when the samples contain $<50\%$ silt and clay (Horowitz 1991).

A slightly different approach is to normalize bulk elemental concentrations by the concentration of a conservative element such as Al, Ti, or Li. In contrast to the methods used for grain size, normalization is performed by dividing the concentration of the potential tracer by the concentration of the conservative element.

The third method commonly used to deal with the transport invariant problem is to incorporate a correction factor into the mixing model (Collins et al. 1998; He and Owens 1995; Russell et al. 2001; Motha et al. 2003, 2004; Juracek and Ziegler 2009). This approach has been widely used to account for the effects of both grain size and organic matter. However, Mukundan et al. (2012) point out that the relationship between a specific fingerprinting parameter and grain size and/or organic matter content may vary between the other parameters used in the composite fingerprint; thus, the incorporation of a single, universally applicable correction factor into the model may not be appropriate. In addition, it has been argued that the use of multiple correction factors, such as one for grain size and one for organic matter, may result in over correction of the parameter values, a problem which is difficult to test (Mukundan et al. 2012).

Once the geochemical parameters that exhibit non-conservative behavior have been removed from the list of potential fingerprints, a statistical test is generally used to identify geochemical properties that are good at discriminating between sediment from various sources. The most commonly used statistical method is the Kruskal-Wallis H-test (Collins et al. 1998, 2001; Walling et al. 1999), but a wide range of other methods have also been applied, including the Mann-Whitney U-test (Carter et al. 2003; Porto et al. 2005), the Wilcoxon rank-sum test (Juracek and Ziegler 2009), and the Tukey test (Motha et al. 2003). A subset of parameters identified during this step is then selected to define the fingerprint that is assumed to represent the optimum combination of parameters for discriminating between the sediment sources or source types (Walling and Woodward 1995; Collins et al. 1998; Mukundan et al. 2012). This last step often relies on the use of a step-wise discriminant function analysis (e.g., Evrard et al. 2013; Miller et al. 2013), although other data reduction techniques (e.g., Principle Component Analysis) have also been used.

2.3.4 Inverse/Unmixing Models

2.3.4.1 Derivation of the Inverse/Unmixing Models

Early work by Yu and Oldfield (1989, 1993) and Collins et al. (1997a, b) was particularly instrumental in defining the general fingerprinting approach most often used today. It can be viewed as a process in which the composite fingerprint created for the sediment sources is compared to the river sediments using an inverse/unmixing model to unravel the relative amount of sediment from each source that comprises the river sediment of interest.

Mathematically, constraints on the mixing model require that (1) each source type contributes some sediment to the mixture, and thus the proportions (x_j, $j = 1, 2, \ldots, n$), derived from n individual source areas must be non-negative ($0 \leq x_j$), and (2) the contributions from all source areas must equal unity, i.e.:

$$\sum_{j=1}^{n} x_j = x_1 + x_2 + \cdots + x_n = 1. \tag{2.3}$$

Three significant factors may lead to situations where this latter assumption of linear additivity in property values is not fully achieved. First, analytical errors may be associated with the characterization of the measured geochemical parameters. These errors are typically on the order of $\pm 5\,\%$, and in most instances do not pose a significant issue. Second, an important sediment source may not have been recognized or sampled. The failure to characterize a significant source primarily occurs when dealing with large basins composed of a large number of sediment sources (geological units, soils types, or land-use/land-cover categories). Third, the tracer(s) may have exhibited non-conservative behavior either during transport, or as a result of diagenetic alterations following deposition (Walden et al. 1997; Rowan et al. 2012).

Assuming that a particular tracer has been established as comprising part of a fingerprint for the n sources, the downstream mixture of this particular tracer is represented by

$$\sum_{j=1}^{n} a_j x_j = a_1 x_1 + a_2 x_2 + \cdots + a_n x_n \tag{2.4}$$

where a_j represents the measurement of the tracer within the jth source area.

Initial studies used the mean or median of the data points from each source in the fingerprint. Collins et al. (2010a), for example, noted that the "use of the mean concentration value to represent a particular source can be justified as being physically realistic since the sediment collected from the catchment outlet inevitably represents a mixture of material mobilized and delivered from numerous locations upstream. As a result, the collection of representative source material samples from a range of locations throughout the catchment and the use of the sample to derive the mean fingerprint property concentrations can be assumed to be analogous to natural sediment mixing during the sediment mobilization and delivery process."

Since the fingerprint will typically be comprised of m tracers, the mixing model results in an $m \times n$ system of linear equations. Each equation represents the contributions from each of the n sources determined on the basis of the measured amount of a tracer in the sediment:

$$
\begin{aligned}
a_{1,1} * x_1 + a_{1,2} * x_2 + \cdots + a_{1,j} * x_j + \cdots + a_{1,n} * x_n &= b_1 \\
a_{2,1} * x_1 + a_{2,2} * x_2 + \cdots + a_{2,j} * x_j + \cdots + a_{2,n} * x_n &= b_2 \\
\vdots \qquad\qquad \vdots \qquad\qquad \vdots \qquad\qquad & \\
a_{i,1} * x_1 + a_{i,2} * x_2 + \cdots + a_{i,j} * x_j + \cdots + a_{i,n} * x_n &= b_i \\
\vdots \qquad\qquad \vdots \qquad\qquad \vdots \qquad\qquad & \\
a_{m,1} * x_1 + a_{m,2} * x_2 + \cdots + a_{m,j} * x_j + \cdots + a_{m,n} * x_n &= b_m
\end{aligned}
\tag{2.5}
$$

In theory, the mixture is exact, but in reality, there will exist some differences (error) between the values of the m measured tracers in the source area, $a_{i,j}$ ($i = 1, 2, \ldots, m, j = 1, 2, \ldots, n$), and the downstream mixture (river sediment), b_i ($i = 1, 2, \ldots, m$). The residual error corresponding to the ith tracer can be determined as follows:

$$
\varepsilon_i = b_i - \sum_{j=1}^{n} a_{i,j} * x_j
\tag{2.6}
$$

for $i = 1, 2, \ldots, m$, where $a_{i,j}$ ($i = 1, 2, \ldots, m, j = 1, 2, \ldots, n$) are measurements of the corresponding ith tracer within the jth source area and b_i is the measurement of the tracer of the ith tracer in the river sediment (mixture).

When the number of utilized tracers exceeds the number of source areas or types within the catchment (as is often the case when using geochemical data), the system of equation (2.5) is over-determined, and a 'solution' is typically obtained using a computational method that optimizes an objective function. This function, subject to the previously noted constraints, estimates a best fit solution to the entire data set (Yu and Oldfield 1989).

There are several ways to obtain a best fit, but in previous studies, the objective function, f, has taken the form of the sum of the relative errors where

$$
f(x_1, \ldots, x_n) = \sum_{i=1}^{m} \left| \frac{\varepsilon_i}{b_i} \right|
\tag{2.7}
$$

(as defined by Yu and Oldfield 1989) or the sum of the squares of the errors (Collins et al. 1997a),

$$
f(x_1, \ldots, x_n) = \sum_{i=1}^{m} \left(\frac{\varepsilon_i}{b_i} \right)^2 = \sum_{i=1}^{m} \left(\frac{b_i - \sum_{j=1}^{n} a_{i,j} x_j}{b_i} \right)^2
\tag{2.8}
$$

Note that the measurements of different tracers are often magnitudes apart; for instance, Cu concentrations ranged from 2.78 to 823 ppm, while Cd ranged from

0.200 to 6.30 ppm within sediments of the Mkabela Basin of South Africa studied by Miller et al. (2013). Thus, the error terms for each tracer in equations (2.7) and (2.8) are normalized by dividing by the amount of the tracer found in the sediment mixture. This insures that the error term of any one tracer does not dominate the objective function.

Ultimately, it is necessary to minimize the function f, (either 2.7 or 2.8), while satisfying the non-negativity constraints on x_j and the unity constraint (2.3). While the error function (2.7) used by Yu and Oldfield (1989) may seem, at first, more intuitive, the mathematical techniques for solving this constrained minimization problem are more arduous than those using the error function given by Eq. 2.8. Since Eq. 2.8 is a quadratic in $(x_1, x_2, \ldots, x_n)$ on a closed convex subset of R^n, the constrained minimization problem is mathematically guaranteed to have a solution. Mathematically, the mixing model would be considered as follows:

$$\begin{aligned} Minimize \ f(x_1, \ldots, x_n) &= \sum_{i=1}^{m} \left(\frac{\varepsilon_i}{b_i}\right)^2 \\ Subject \ to \qquad \sum_{j=1}^{n} x_j &= 1 \\ x_j &\geq 0 \end{aligned}$$

(2.9)

Such problems, known as quadratic programming problems, are well understood. Many mathematical programs, such as MATLAB, *Mathematica*, even Excel, have built in programs to solve these problems. There are other packages that solve the quadratic programming problem as well.

Rowan et al. (2000) solved a different form of objective function based on a variation of the R-value used in regression. They referred to the objective function as an efficiency function, E. This efficiency function, E, is defined as:

$$E(x_1, \ldots, x_n) = 1 - \frac{\sum_{i=1}^{m}(b_i - \sum_{j=1}^{n} a_{i,j}x_j)^2}{\sum_{i=1}^{m}(b_i - \frac{1}{n}\sum_{j=1}^{n} a_{i,j})^2}$$

(2.10)

Instead of being minimized, E was maximized subject to the non-negativity and unity (2.3) constraints.

2.3.4.2 Solving the Optimization Problem

A criticism of using mixing models to determine the relative contribution of sediments from a source is that there may be a number of solutions that are statistically equivalent, particularly where contributions from a given source approach 0 or 100 % (often referred to as the equifinality problem). In other words, similar levels of model performance as measured by an error or efficiency function can be produced by differing sets of source contributions (Collins et al. 2010a; Rowan et al. 2012). In addition, uncertainty in the modeling results may be associated with (1) the inherent variability of the fingerprint within the source materials, (2) the source material sampling density, (3) analytical errors, and (4) changes in sediment characteristics during particle entrainment, transport and deposition which may significantly influence the chemical and physical nature of the sampled deposits, such as grain size and organic matter content (Small et al. 2004).

Without the benefit of formalized numerical optimization techniques, initial attempts at solving the constrained optimization problems (e.g., by Rowan et al. 2000; Jenkins et al. 2002; Phillips and Gregg 2003) involved pushing various combinations of $(x_0, x_1, \ldots, x_n)$ that satisfied the non-negativity function and the unity condition through the objective function to find the optimal solutions. All values of the objective function were then recorded and compared to determine the optimum value and those combinations of proportions which yielded this optimal value.

As an example, to solve the mixing problem using the efficiency function (2.10), Rowan et al. (2000) created all possible combinations of proportions, $(x_1, x_2, \ldots, x_n)$, by generating all n-tuples (over 300,000 in all for 5 sources) differing by increments of $\Delta x = 0.02$. They then substituted them into the efficiency function E. By plotting E vs x_i, Rowan et al. (2000) were able to determine for each source a range of proportions that would generate an efficiency above a certain tolerance.

Jenkins et al. (2002) applied the same method to terrestrial and marine sediments to determine sediment provenance using mineral magnetic properties as a fingerprint. The approach has also been used in disciplines other than geomorphology. Phillips and Gregg (2003), for example, applied the method to determine the structure of food-webs using stable isotopes as a fingerprint of food sources.

While, this technique allows one to generate numerous values of the objective function, it is severely limited by the size of memory necessary to record all possible values of f with $(x_1, x_2, \ldots, x_n)$ for large numbers of sources beyond increments of $\Delta x \leq 0.01$. Later attempts at solving the mixing model recognized the constrained optimization problem as a standard quadratic programming problem and used available packages to solve it, as mention in Sect. 2.3.4.1.

2.3.4.3 Modifications to the Mixing Model

Since the late 1990s, a number of modifications have been made to mixing models to improve upon their overall effectiveness. One of the first modified the objective function in the mixing model (2.8) to account for differences in grain size and organic matter content between the source area sediments and the river sediments (Collins et al. 1997a, 2001). Later, Collins et al. (2010a) made two additional modifications to the objective function. First, they added a 'within source variability' weight. They found the use of this weighting parameter gave smaller ranges of possible source contributions when calculated using a Monte Carlo method (discussed below). Second, they introduced a tracer discriminatory weight to 'reflect the tracer discriminatory power.' This weighting factor takes into account the relative ability of a specific fingerprinting parameter to differentiate between the various sediment sources.

When the above factors are incorporated into the objective function (2.8), it exhibits the following form:

$$f(x_1, \ldots, x_n) = \sum_{i=1}^{m} \left(\frac{b_i - \sum_{j=1}^{n} a_{i,j} * ps_j * om_j * ws_{i,j} * x_j}{b_i} \right)^2 * W_i \qquad (2.11)$$

where ps_j and om_j are the particle size and organic matter weights, respectively, for the jth source; $ws_{i,j}$ is the within source weight for the ith tracer within the jth source; and W_i is the discriminatory weight for the ith tracer. Specific definitions for the correction factors that have been added by Collins and his colleagues to the model are provided Table 2.2.

Another form of modification that has recently received considerable attention is related to Bayesian statistics. Walling and Collins (2005), and Collins et al. (2013), for example, modified the approach to incorporate prior knowledge about river processes into the model. More specifically, they limited the potential contribution from one of the sources within the model by noting, in this case, that the contribution of sediment to the channel by bank erosion could be no more than 50 %. Thus, they imposed an additional constraint on the model. That is, in addition to the unity constraint (2.3), when x_1 represents the proportion of the mixture due to channel bank erosion, it must adhere to the following constraints: $0 \leq x_1 \leq 0.5$ and $0 \leq x_j$ for $j = 2, 3, \ldots, n$.

2.3.4.4 Handling Uncertainty of Tracer Data: The Monte Carlo Method

A significant source of uncertainty in the fingerprinting approach is the inherent variability of a fingerprinting parameter within the source and river sediments. Take, for example, a source area defined by its land-cover such as forest vegetation. Forested areas may cover multiple geological units or soil types, each possessing a unique mineralogical and geochemical set of characteristics. Thus, the variability inherent in the collected geochemical data is likely to be relatively high. When combined with

Table 2.2 Definition of correction/weighting factors used in inverse models (Collins et al. 1997a, 2001, 2010a)

Correction/weighting factor	Definition
Particle size correction factor	Ratio of specific surface area measured for a given source to the average of the surface area from all the sources
Organic matter correction factor	Ratio of the organic carbon content within a given source to the average organic carbon content of all the sources
Within source variability weight	The inverse of the standard deviation of a parameter for all samples from the source
Tracer discriminatory weight	Calculated for each tracer as the ratio of the percentage of source type samples classified correctly for that tracer to the minimum of all such measurements determined during the discriminant analysis used to identify the fingerprint

the fact that the number of samples collected to characterize the sediment sources are generally limited because of financial and time constraints, the mean value of the fingerprinting parameter calculated from the sample data from any source may not necessarily represent the true mean of each tracer within the source. Thus, using the average of the data points of all samples collected within the source for each tracer elicits an error of unknown magnitude. This error is exacerbated by the fact that the fingerprint is comprised of several tracers (Collins et al. 2010a).

In order to reduce and quantify the uncertainty in mixing/unmixing models related to this inherent variability in the source area data, recent studies have explored the use of a Monte Carlo sampling framework (Small et al. 2004; Collins et al. 2010a, 2012). Validation of these approaches using constructed laboratory mixtures of source materials and synthetic data show the methods hold considerable promise (Small et al. 2004).

As an example, Small et al. (2002, 2004) used the sample data from each source to create a probability distribution to estimate the mean value for each tracer within the source. The estimated means of all tracers within all sources are then used as parameters in the objective function which is minimized to solve for the proportions. This process is repeated numerous times (on order of several thousands) until there are sufficient results to estimate confidence intervals (e.g., 95 %). As described in more detail below, Rowan et al. (2012) used this method to determine the effects land use management practices have on algae blooms.

Collins et al. (2010a) used a similar approach in which a goodness of fit function, which they refer to as a relative mean error (RME), was assumed to measure the robustness of the optimized solutions of the mixing model:

$$RME = 1 - \left[\frac{1}{m} \sum_{i=1}^{m} \left(\frac{b_i - \sum_{j=1}^{n} a_{i,j} * ps_j * om_j * ws_{i,j} * x_j}{b_i} \right)^2 * W_i \right] \quad (2.12)$$

Subsequent studies by Collins et al. (2012, 2013) modified the approach by using the median values instead of mean values for the tracer parameter within each source. The median values were estimated on the basis of a probability density function. They also use an estimated frequency-weighted average median, initially introduced by Collins et al. (2012), in which $R = \sum_{i=1}^{n} v_i Fr_i$ where n is the number of intervals for the predicted deviate relative contribution, scaled between 0 and 1, and v_i and Fr_i are the mid-value and the frequency for the ith interval, respectively.

A question that arises in the use of the fingerprinting approach is how results obtained from using mean source and river sediment values compare to the output generated using the Monte Carlo approach. While additional investigations are needed to answer this question, Fig. 2.6 compares the result generated using only mean values (arrows) to the results obtained using the objective function (2-norm error squared) and confidence interval of 95 % using the RME defined above (2.12) for six samples collected from a wetland core within the Mkabela catchment of South Africa. Grain size and organic matter correction factors were not included in the analysis. While differences exist, the results for these specific samples are

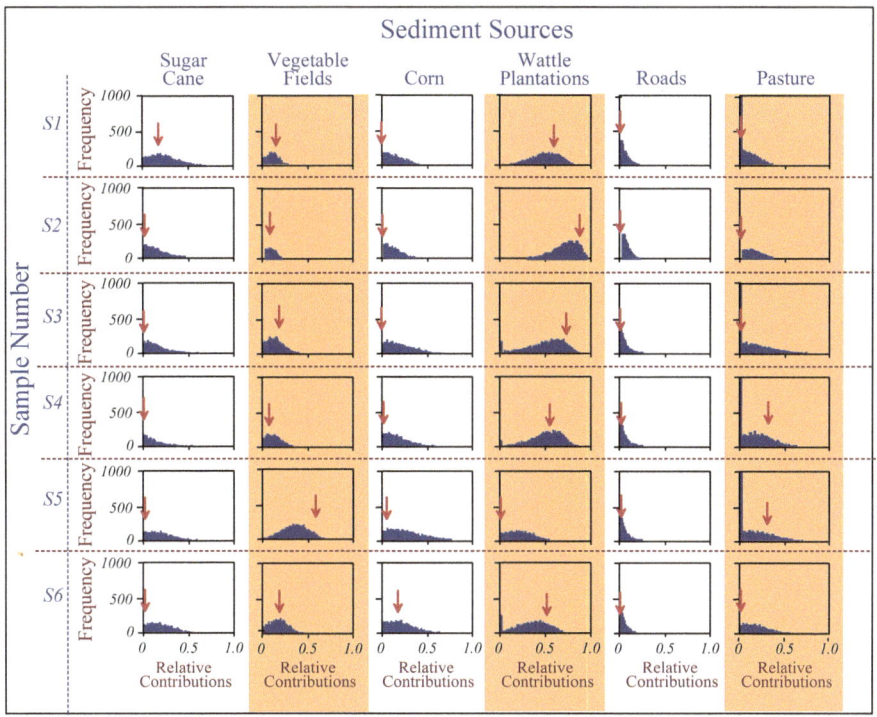

Fig. 2.6 Comparison of relative contributions of sediment estimated using a Monte Carlo approach (frequency diagrams) versus using only source means for wetland core samples collected within the Mkabela Basin, South Africa. *Red arrows* represent average contribution derived using only source means. In general, results are similar for these data, but the Monte Carlo method allows for an assessment of result uncertainty

generally comparable. The primary difference is, of course, that uncertainty can be accessed using the Monte Carlo method.

2.4 Applications

Thus far we have focused on the methods of using geochemical fingerprinting to determine the provenance of river sediments, and the assumptions and uncertainties inherent in the produced results. We will now examine a few individual studies in more detail. The review is not intended to be exhaustive of the wide range of investigations that have utilized geochemical fingerprinting, but rather to provide an overview of some the types of problems that may be addressed using the fingerprinting approach.

A number of early geochemical fingerprinting studies were aimed at determining the source of contemporary sediments in rivers by sampling suspended sediments during flood events (e.g., Peart and Walling 1986; Walling and Wood-ward 1992, 1995; Walling et al. 1993; Wood 1978; Collins et al. 1997a,b, 1998, 2001). Collins et al. (2001), for example, collected 65 samples from 13 floods during two wet periods between 1997 and 1999 within the upper Kaleya catchment of southern Zambia. The objective was to determine the predominant contributions of sediment from four sources including bush grazed lands, commercially cultivated lands, communally cultivated lands, and channel banks and gullies. As is now common, the collected sediment samples were analyzed for a range of properties that respond to different environmental controls, an approach that is expected to lead to better source area sediment discrimination because the individual tracers will exhibit a higher degree of independence (Walling et al. 1993; Collins et al. 1997b). Collins et al. (2001) chose to analyze sediments for pyrophosphate-dithionite extractable trace metals (Al, Fe, Mn), acid extractable metals and metalloids (As. As, Cd, Co, Cr, Cu, Fe, Mn, Ni, Pb, Sb, Sn, Sr, Zn), base cations (C, K, Mg, Na), organic matter (C, N), radionuclides (^{137}Cs, ^{210}Pb$_{ex}$, ^{226}Ra) and total P. Source contributions were determined using the unmixing model described by Collins et al. (1997a), which included correction factors for grain size and organic matter as well as a weighting factor to account for differing degrees of analytical precision between the tracers. They found that surface soils in communally cultivated lands were the predominant sediment source. However, there were minor variations in source contributions between the samples collected during an individual flood (Fig. 2.7a). More importantly, because the samples were collected during different discharge and sediment load conditions, attempts to decipher source contributions during an entire flood (or longer time periods) needed to consider the sediment load at the time the suspended sediment samples were collected. By considering load, the contributions associated with samples characterized by higher sediment loads are given more weight than samples characterized by lower sediment loads. Mathematically, a load-weighted mean source contribution (P_{sw}) can be calculated for any source(s) for any given time period (Walling et al. 1999) using the following equation:

$$P_{sw} = \sum_{s=1}^{n} P_{sx} \frac{L_x}{L_t} \qquad (2.13)$$

where L_x is the instantaneous suspended sediment load at the time of sample collection, L_t is the sum of the instantaneous sediment loads for all samples collected during the time period of interest (e.g., a flood or season), and P_{sx} is the percentage contribution from a specific sediment source, s, to the sediment sample, x. When applied to an individual flood, the weighted mean source contribution provides information on the source of the sediments transported past the monitored site during the event. In the case of the upper Kaleya catchment, Collins et al. (2001) found that in addition to the noted intra-flood variability in source contributions, inter-flood and seasonal variations in contributions occurred (Fig. 2.7b).

The use of suspended sediment samples to determine the primary sources of sediment to a channel has decreased in recent years as it is plagued by several difficulties, including (1) the fact that estimates of sediment source within suspended

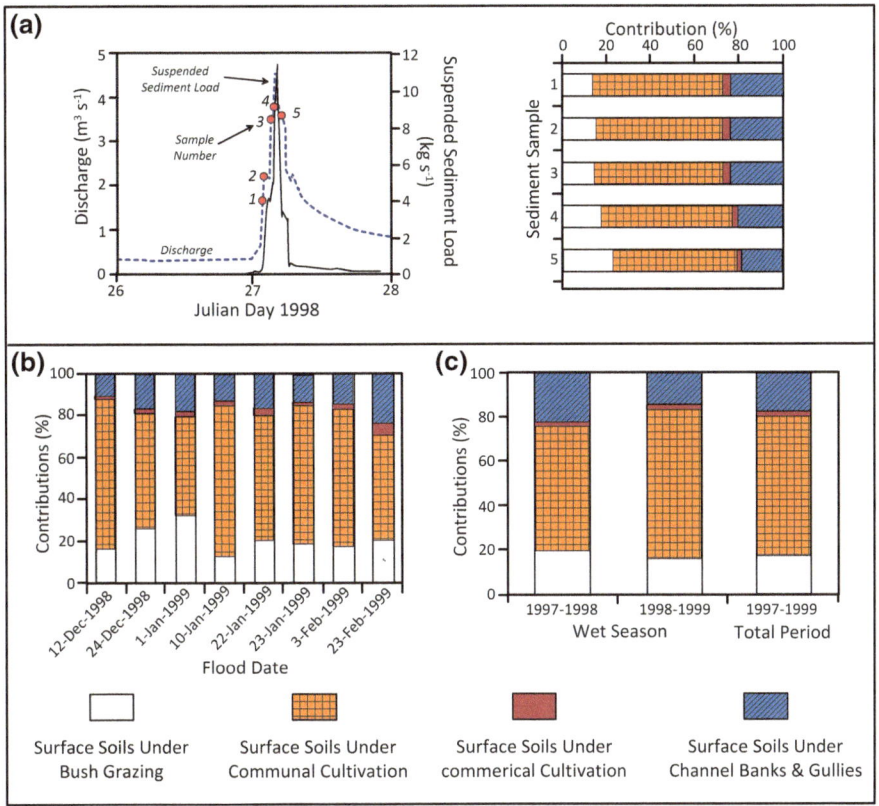

Fig. 2.7 **a** Estimated relative source contributions to suspended sediment sampled during a flood event in 1998 within the upper Kaleya Catchment. Note intra-flood variability in sediment provenance. **b** Load-weighted sediment source contributions for runoff events sampled in 1998 and 1999 (*left*), and calculated for the entire wet periods between 1997 and 1999 (*right*) within the upper Kaleya Catchment (adapted from Collins et al. 2001)

samples are sensitive to the timing of sample collection, (2) the need to collect samples during flood events; (3) the need to collect and process significant volumes of water to obtain enough sediment for geochemical analysis, and (4) the costs and time required for processing a large number of samples to adequately characterize contributions during a protracted time period (e.g., season or year). More recent studies have focused on sediment collected using passive sediment traps/samplers, channel bed sediments, or sediments located at the surface of floodplains.

Collins et al. (2010a), for example, applied geochemical fingerprinting to river sediments obtained from the floodplain surface near the mouth of seven subcatchments of the Somerset Levels in the UK (Fig. 2.8). These samples were assumed to represent sediments eroded during moderate floods that transported the majority of the sediment in the basins. The study area is part of the England Catchment Sensitive Farming Delivery Initiative where Catchment Sensitive Farming Officers are

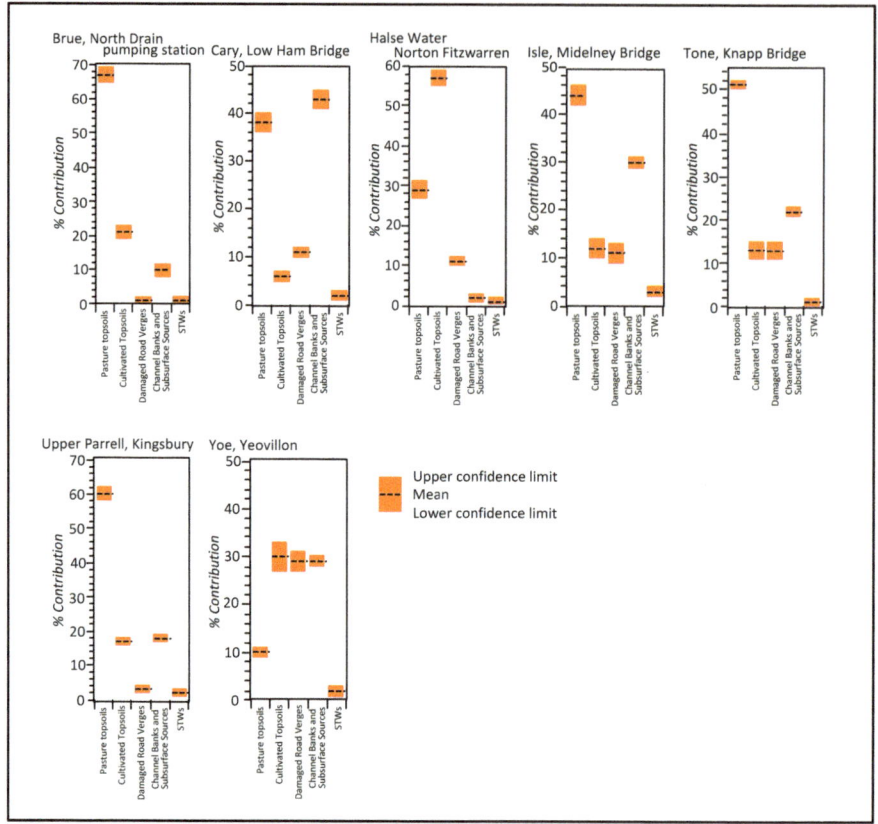

Fig. 2.8 Proportional contribution of sediment to floodplain surface samples collected at the mouth of seven subcatchments in the Somerset Levels, south west UK. Note variability in source contributions between catchments (from Collins et al. 2010a)

responsible for assessing the potential sources and impacts of pollutants on aquatic resources, and for providing advice to stakeholders who are in need of assistance in protecting aquatic environments (Collins et al. 2010b). The goal of the project was to determine the relative contribution of sediment from five sediment sources: pasture lands, cultivated topsoil, damaged road verges, channel banks and other subsurface sources (e.g., gullies), and sewage treatment wastes (a point source). The fingerprint was developed using the procedures outlined by Collins et al. (1997a) from a suite of 40 geochemical parameters analyzed for each of the source area samples. Inverse modeling relied on a Monte Carlo approach and the modified optimization method descibed above which included correction factors for grain size and organic matter content as well as weightings for tracer discriminatory ability and within-source variability of the individual tracers. The generated results were provided in terms of the mean contribution from the defined sources and their 95 % confidence limits. Figure 2.8 shows that the contribution from each of the sediment sources varied significantly between the seven subcatchments. The observed variations in source area inputs to the river not only reflected the proportion of the basins covered by the sediment source, but also a host of other factors that controlled the erosion of the source sediments. Collins et al. (2010a) argued, for instance, that sediment inputs from bank erosion were probably influenced by such factors as "channel morphology and density, river bank dimensions, and riparian land use pressures". Regardless of the controls, the data clearly indicated that sediment mitigation strategies will need to be tailored to individual subcatchments, rather than using a one-method fits all approach.

Evrard et al. (2013) utilized geochemical fingerprinting (and an alternative, diffuse reflectance infrared Fourier Transform spectroscopy method) to assess the predominant source of sediments to tropical rivers in central Mexico underlain by different soil types. In this case, the goal was to determine whether the sediments were primarily produced by gully erosion or sheet erosion. As they were attempting to decipher sediments eroded from surface and subsurface sites, they relied on fallout radionuclides (^{137}Cs and ^{210}Pb$_{ex}$) and biogenic elements (C, N) as fingerprints. These parameters are known to differ significantly between surface and subsurface sources (as will be discussed in the next chapter). The analyzed river sediments were obtained in 2009 from the channel bed, and were assumed to represent average contributions to the river during the rainy season. Evrard et al. (2013) found that within the Huertitas subcatchment dominated by Acrisols the majority of the sediment (between 88 and 98 %) was derived from gullies. The amount of sediment derived from croplands by sheet erosion decreased during the rainy season, possibly as a result of increased vegetation cover that helped stabilize the soil surface. In contrast, the majority of the sediment (50–85 %) within the Andisol dominated La Cortina catchment was derived from the surface of croplands. Within the Potrerillos catchment, characterized by both Acrisols and Andisols, contributions of sediment from gullies and croplands were highly variable between storms. Gullies, for example, generated between 5 and 86 % of the sediment, while the sheet erosion of rangelands generated between 14 and 95 % of the sediment. However, when combined with other data, it appears that fine-sediment delivered to the Cointzio reservoir was primarily derived from gullies developed in Acrisols, even where Acrisols covered small areas of the basin (<0.5 %). Thus, they suggested that sediment mitigation efforts should focus on stabilizing gully networks.

While the three analyses described above focused on contemporary sediments, other studies have combined their investigation of contemporary sediment sources with an analysis of the changes in sediment provenance through time. These studies typically focus on cores taken of semi-continuously accumulated sediments on floodplains (e.g., overbank deposits) or within reservoirs, lakes, or wetlands that receive sediments from the upstream drainage network. Such fingerprinting studies can be extremely useful from a management perspective in that the sediment cores can be dated, allowing changes in sediment provenance to be linked to (1) an analysis of sediment accumulation rates and the changes in those rates through time, (2) temporal changes in the concentration of various contaminants including trace metals and nutrients within the system, and (3) alterations in land-use/land-cover and other human activities within the catchment. Thus, a link can be made between sediment-contaminant source, sediment influx rates to the river, and anthropogenic activities.

Rowan et al. (2012), for example, applied the geochemical fingerprinting approach to cores extracted from a lake—Llyn Tegid located in the headwaters of the River Dee, Wales—to assess the potential cause(s) of recent cyanobacteria blooms. They found by dating the cores (using ^{210}Pb and ^{137}Cs methods) that the sedimentation rates within the lake were generally constant during the first half of the 20th century (Fig. 2.9). However, sediment accumulation rates began to progressively increase

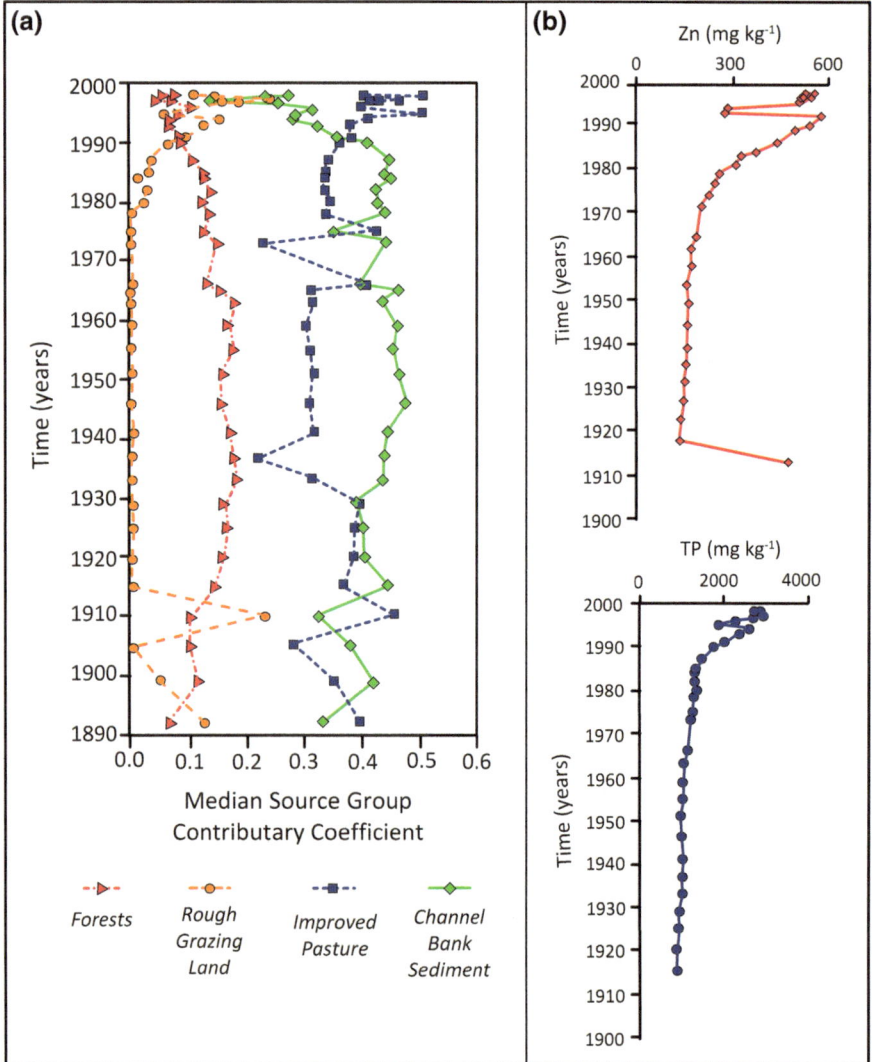

Fig. 2.9 a Variations in sediment source contributions through time as determined from the incremental sampling and analysis of a core from Llyn Tegid, a lake in the Snowdonia National Park, Wales. **b** Downcore trends in total phosphorus and Zn (modified from Rowan et al. 2012)

after about 1950, with the most significant increases occurring in the 1980s and 1990s. The largest increases in sedimentation in 1995 were correlated with increases in sediment-associated trace metal and nutrient (primarily total P) concentrations and blue-green algae blooms. In addition, Rowan et al. (2012) were able to show using an unmixing model described by Franks and Rowan (2000) that the majority of the sediments were derived from improved grazing lands (i.e., lands subjected to improved management practices) and channel bank erosion throughout the period of record. However, the observed increase in sediment accumulation rates was closely linked to an increased supply of sediment from improved pastures in the late 1980s. Improved grazing lands subsequently became the predominant source throughout the 1990's during which they supplied about 50 % of the total sediment to the lake. Increases in sedimentation rates were also associated with increased sediment yields from rough grazing areas (Fig. 2.9). The accelerated flux of sediment from both the improved and rough grazing areas was consistent with an intensification of agricultural activity in the basin in the 1980s. Rowan et al. (2012) concluded that the algae blooms were strongly influenced by the increased influx of sediment-associated nutrients from grazing lands, particularly improved grazing lands which were actively managed. Following the establishment of algal populations in the lake, simulation modeling of phytoplankton dynamics suggested that even a 50 % reduction in total P input would not eradicate the algae problem because several factors will work to maintain them, including a pool of P in the lake bed sediments, flow augmentation from the Trywryn and impoverished littoral macrophyte assemblages.

By analyzing multiple cores from wetlands and reservoirs located along the axial drainage of the Mkabela Basin within KwaZulu-Natal Midlands of eastern South Africa, Miller et al. (2013) demonstrated that it was possible to decipher both the temporal changes in provenance to the drainage network at any given site and spatial changes in sediment provenance along the valley system at any given time. Their study design also differed from many previous efforts in that source areas were defined and analyzed using two spatial classification systems. One system examined sediment source by soil type, whereas the other defined sediment source according to five land-use categories including pastures, pine forests, sugar cane fields, vegetable plots, and wattle groves. Geochemical fingerprints for the sediment sources were determined for both classification systems using the same set of sample data, but which had been stratified differently. Results from an unmixing model, based on the optimization method described by Rowan et al. (2000), showed that silt- and clay-rich layers found within wetland and reservoir deposits of the upper and upper to mid-portions of the basin were derived from the erosion of fine-grained, valley bottom soils frequently utilized as vegetable fields (Fig. 2.10). These sediments also exhibited elevated concentrations of Cu and Zn, presumably from the use of fertilizers. In contrast, coarser-grained sediments were primarily derived from the erosion of sandier hillslope soils, extensively utilized for sugar cane, during relatively high magnitude runoff events that were capable of transporting sand-sized sediment off the slopes. Thus, the combined data showed that the complex interactions between runoff, soil type, and land use (among other factors) created temporal and spatial variations in sediment provenance. Moreover, downstream contrasts in sediment

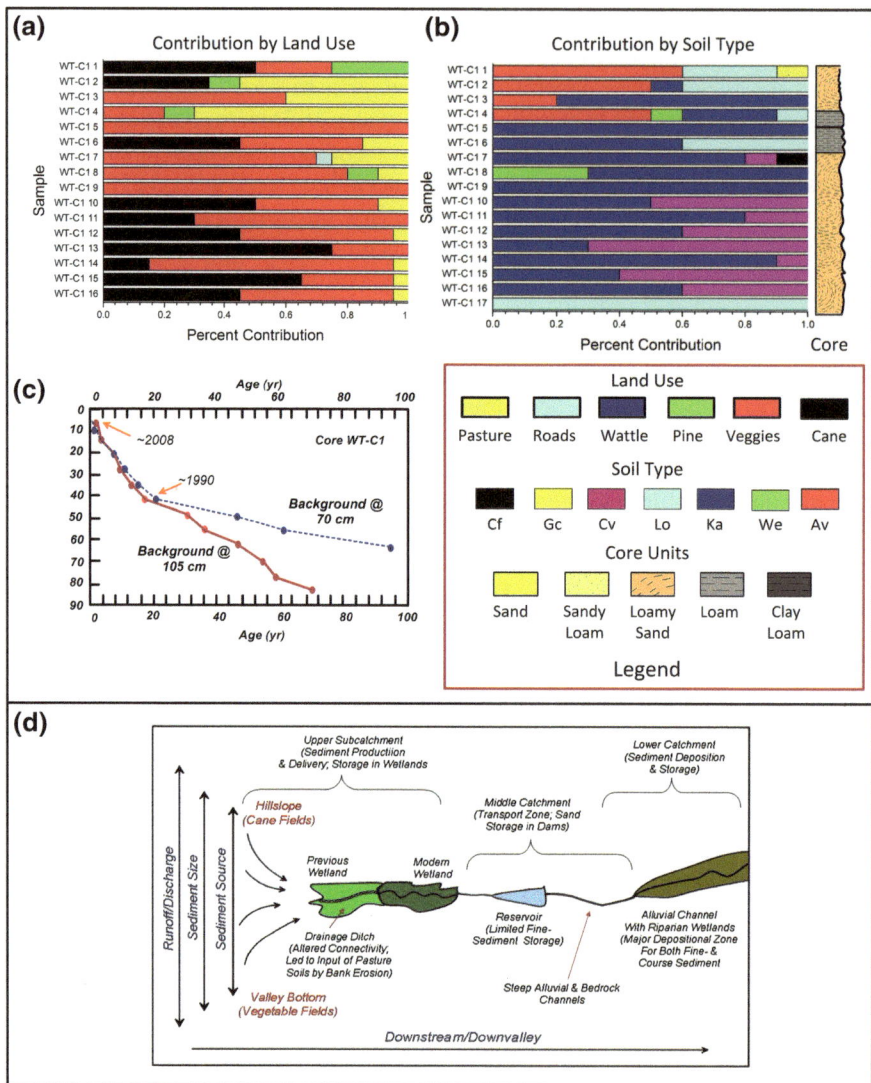

Fig. 2.10 Estimated contributions of sediment to a wetland from **a** delineated land-use categories and **b** soil types. **c** Estimated age of wetland sediments using two different models. Sedimentation rates increase after 1990 as a results of a constructed drainage ditch that allowed sediment to be transported to the wetland. **d** Schematic diagram of predominant sediment sources and the variations in sediment size and source with runoff magnitude (adapted from Miller et al. 2013)

source contributions, combined with observed changes in Cu and Zn within the cores, suggested that sediment export from upper to lower catchment areas was limited until the early 1990s, in part because the lower parts of the watershed were hydrologically disconnected from the upper catchment wetlands during low- to moderate flood events (Fig. 2.10). The construction of a drainage ditch through an upstream wetland altered the hydrologic connectivity of the catchment, allowing sediment and sediment-associated nutrients to be transported from the headwaters to the lower basin, a process that also increased downstream sedimentation rates as determined by dating a sediment core. From an applied perspective, the results of the study showed that the positive benefits of controlling sediment/nutrient exports from the catchment by means of upland based best management practices were partly negated by modifying the axial drainage system.

The above discussion may suggest that geochemical fingerprinting to assess sediment provenance is a straightforward process that can be utilized by a relatively well-defined methodology. In reality, each catchment exhibits a unique set of characteristics requiring investigators to modify the approach to fit the catchments hydrologic, geomorphic and climatic setting. Nonetheless, the rapidly increasing volume of literature on the topic shows that fingerprinting can provide important insights into the temporal and spatial changes in sediment provenance for specific grain size fractions. Moreover, it is apparent that geochemical fingerprinting can be combined with other forms of geomorphic, hydrologic, and geochemical tracing to more fully address a wide range of sediment related problems within a watershed.

2.5 Use of Geochemical Fingerprinting as a Management Tool

There is little debate that sediment source identification is a fundamental requirement for the effective mitigation of diffuse sediment and sediment-born contaminant inflows to rivers and other aquatic environments. However, the use of fingerprinting techniques to decipher the source and dispersal of non-point source contaminants, including sediment, at the catchment scale has yet to be extensively utilized by land-use managers or regulatory agencies. In fact, Mukundan et al. (2012) found that only one state in the U.S. (Minnesota) was using geochemical fingerprinting as part of a defined management strategy, and, to the best of our knowledge, it remains the only state as of 2014. Given the nature of recent legislation in many developed countries, and the need for a sound understanding of the predominant sources of sediment to rivers, it seems likely that geochemical fingerprinting will more extensively be used in the future for management or regulatory purposes. Actually, the potential benefits and difficulties of transforming geochemical fingerprinting from a research to a management tool is currently being explored. Mukundan et al. (2012), for example, examined the use of geochemical fingerprinting for the establishment of total maximum daily loads (TMDLs) in the U.S. TMDLs, which states must define for impaired waters as part of a management strategy, represent the maximum amount of a given pollutant, in this case sediment, that the water body may receive without violating water-quality standards. A key component of the USEPA's organizational framework for establishing TMDLs is contaminant source assessment; thus, it would seem that geochemical fingerprinting could (and perhaps should) be incorporated into the framework for establishing TMDLs. The analysis by

Mukundan et al. (2012) shows, however, that the transformation of geochemical fingerprinting from the research to the management/regulatory realm is not as straight forward as one might think, and has been slowed by several factors.

1. *Many land-use managers and regulatory personal are unfamiliar with the specifics of the approach.*

   In comparison with many traditional approaches, land-use managers/regulators may have a relatively poor understanding of geochemical fingerprinting techniques, and therefore do not necessarily understand the benefits of incorporating the methods into their management framework. As Mukundan et al. (2012) point out, they are likely to question the practicality of applying the approach, especially with regards to the cost and time required for the analysis, the spatial scale to which it can be applied, the type of geochemical analyses that must be carried out, and the likelihood of obtaining meaningful results. Moreover, the benefits of applying geochemical fingerprinting methods as part of a larger management strategy may be unclear. Sediment budgeting, for example, is a relatively well known method of conceptualizing the sediment dispersal system. However, defining the terms within a sediment budget by monitoring sediment loads, quantifying upland erosion rates, or through the use of empirically or physically based models have proven problematic. In contrast, fingerprinting has been shown to be an effective approach to estimate the relative contributions of sediment from defined sources at the catchment scale, and can be applied within a relatively short time frame. It allows, then, for the targeting of the primary source areas or types to reduce sediments loads within the catchment. In addition, mixing model results may be combined with upland erosion and downstream sediment load data to determine: (1) the fraction of the sediment load generated from each sediment source that exists within the basin, and (2) the fraction of sediment eroded from upland areas that is deposited and stored within channels and floodplains (Mukundan et al. 2012). Similarly, geochemical fingerprinting can be linked to the results obtained from watershed models to determine the export and storage of sediment from individual sources. Thus, fingerprinting can be incorporated into the traditional budgeting or modeling approach to refine and greatly improve upon its overall results, a fact that is not always recognized. It is also important to note that the conversion of relative source contributions to estimates of mass sediment transport may allow for a comparison with the more traditional methods of measuring sediment inputs to rivers from the defined sources. For example, rates of bank erosion determined through repeated channel surveys may be compared to estimates of bank erosion influx made by fingerprinting techniques. In doing so, fingerprinting serves as a way to assess the uncertainty inherent in the outcomes of the traditional methods (Mukundan et al. 2012).

2. *Currently, a well-defined set of procedures are lacking.*

   Most analyses performed within a regulatory framework are governed by a well-defined set of operating procedures that are accepted by the 'scientific community' and that lead to accurate and reproducible results. The intent is to produce results that can withstand the rigors of both scientific and legal review. At the present time, standardized procedures that can be used on a routine basis for geo-chemical fingerprinting are generally lacking (Mukundan et al. 2012). Particular aspects of the approach that require some form of guidance and/or standardization are many, including the number of samples required for source area character-ization and the methods used to collect them, the type of river sediments that should be sampled and analyzed to assess sediment source contributions for a given timeframe, the approach(es) that should be used to alleviate the problems of hydraulic sorting and other processes that modify the sediment as it is dis-persed through the system, the quantitative approach that should be followed to define the most discriminating fingerprints, the nature of the mixing mod-els to be utilized, and the methods through which uncertainty in the modeling results can be characterized and assessed, to mentioned just a few. While it can be argued that such procedures should be developed by the agency(ies) that intend on using the approach, the transformation of fingerprinting methods from research to management tool will require significant input from the scientific commu-nity.

3. *Geochemical fingerprinting of diffuse sources at the catchment scale has yet to be completely accepted by the scientific community.*
   The results from geochemical fingerprint studies conducted on diffuse pollu-tion sources continues to be met with skepticism by some highly-respected geomorphologists and other environmental scientists. A primary issue for some is the inability of the conventional mixing models to quantitatively assess the uncertainty of the generated results. Recent refinements in the models to reduce, assess, and quantify the uncertainty in their results is likely to alleviate much of this concern in the future. For other scientists, the concern rests on the assumptions that form the foundation for inverse modeling. For example, in many instances geochemical fingerprinting documents the ultimate source of the sediment, but not necessarily the most recent source as the sediment may have been eroded and transported intermittently along the drainage network before being sampled. Additional studies will need to be performed to assess the degree to which the assumptions inherent in fingerprinting are met.

References

Abu-Zeid M, Baghdady A, El-Etr H (2001) Textural attributes, mineralogy and provenance of sand dune fields in the greater Al Ain area, United Arab Emirates. J Arid Environ 48:475–499

Alt-Epping U, Stuut J, Hebbeln D, Schneider R (2009) Variations in sediment provenance during the past 3000 years off the Tagus River, Portugal. Mar Geol 261:82–91

Amidon W, Burbank D, Gehrels G (2005) Construction of detrital mineral populations: insights from mixing of U-Pb zircon ages in Himalayan rivers. Basin Res 17:463–485

Armstrong A, Maher B, Quinton J (2010) Enhancing the magnetism of soil: the answer to soil tracing? EGU general assembly 2010. Geophys Res Abstr 12:EGU2010-2805

Armstrong J, Kemp P, Kennedy G, Ladle M, Milner N (2003) Habitat requirements of Atlantic Salmon and Brown trout in rivers and streams. Fish Res 62:143–170

Arribas J, Critelli S, Le Pera E, Tortosa A (2000) Composition of modem stream sand derived from a mixture of sedimentary and metamorphic source rocks, Henares River, Central Spain. Sediment Geol 133:27–48

Basu A, Molinaroli E (1991) Reliability and application of detrital opaque Fe-Ti oxide minerals in provenance determination. Geol Soc Lond Spec Publ 57:55–65

Benedetti M, Raber M, Smith M, Leonard L (2006) Mineralogical indicators of alluvial sediment sources in the Cape Fear River basin, North Carolina. Phys Geogr 27:258–281

Bernet M, Bassett K (2005) Provenance analysis by single-quartz-grain SEM-CL/optical microscopy. J Sediment Res 75:492–500

Billheimer D (2001) Compositional receptor modeling. Environmetrics 12:451–467

Bo T, Fenoglio S, Malacarne G, Pessino M, Sgariboldi F (2007) Effects of clogging on stream macroinvertebrates: an experimental approach. Limnol—Ecol Manag Inland Waters 37:186–192

Bottrill L, Walling D, Leeks G (2000) Using recent overbank deposits to investigate contemporary sediment sources in larger river basins. Tracers in geomorphology. Wiley, Chichester

Bravo-Espinoza M, Mendoza ME, Medina-Orozco L, Prat C, Garcia-Oliva F, Lopez-Granados E (2009) Runoff, soil loss and nutrient depletion under traditional and alternative cropping systems in the Transmexican volcanic belt, Central Mexico. Land Degrad Dev 20:640–653

Caitcheon G (1998) The significance of various sediment magnetic mineral fractions for tracing sediment sources in Killimicat creek. Catena 32:131–142

Cardona JPM, Mas JMG, Bellon AS, Domiinguez-Bella S, Lopez JM (2005) Surface textures of heavy mineral grains: a new contribution to provenance studies. Sediment Geol 174:223–235

Carter J, Owens P, Walling D, Leeks G (2003) Fingerprinting suspended sediment sources in a large urban river system. Sci Total Environ 314–316:513–534

Collins AL (1995) The use of composite fingerprints for tracing the sources of suspended sediment in river basins. PhD thesis

Collins AL, Walling DE (2002) Selecting fingerprinting properties for discriminating potential suspended sediment sources in river basins. J Hydrol 261:218–244

Collins AL, Walling DE (2004) Documenting catchment suspended sediment sources: problems, approaches and prospects. Prog Phys Geogr 28:159–196

Collins AL, Walling D, Leeks G (1996) Composite fingerprinting of the spatial source of fluvial suspended sediment: a case study of the Exe and Severn river basins, United Kingdom. Geomorphol: Relief, Process Environ 2:41–53

Collins A, Walling D, Leeks G (1997a) Use of the geochemical record preserved in floodplain deposits to reconstruct recent changes in river basin sediment sources. Geomorphology 19:151–167

Collins A, Walling D, Leeks G (1997b) Source type ascription for fluvial suspended sediment based on a quantitative composite fingerprinting technique. Catena 29:1–27

Collins AL, Walling DE, Leeks GJL (1998) Use of composite fingerprints to determine the provenance of the contemporary suspended sediment load transported by rivers. Earth Surf Proc Land 23:31–52

Collins AL, Walling DE, Leeks GJL (2005) Storage of fine-grained sediment and associated contaminants within the channels of lowland permeable catchments in the UK. In: Walling DE, Horowitz AJ (eds) Sediment budgets 1. International Association of Hydrological Sciences Publication No 291. IAHS Press, Wallingford, pp 259–268

Collins AL, Walling DE, Sichingabula HM, Leeks GJL (2001) Suspended sediment source fingerprinting in a small tropical catchment and some management implications. Appl Geogr 21:387–412

Collins A, Walling D, Webb L, King P (2009) Particulate organic carbon sources and delivery to river channels in the Somerset levels ECSFDI priority catchment, southwest UK. Int J River Basin Manag 7:277–291

Collins A, Walling D, Webb L, King P (2010a) Apportioning catchment scale sediment sources using a modified composite fingerprinting technique incorporating property weightings and prior information. Geoderma 155:249–261

Collins A, Walling D, Stroud R, Robson M, Peet L (2010b) Assessing damaged road verges as a suspended sediment source in the Hampshire Avon catchment, southern United Kingdom. Hydrol Process 24:1106–1122

Collins AL, Walling DE, McMellin GK, Zhang Y, Gray J, McGonigle D, Cherrington R (2010c) A preliminary investigation of the efficacy of riparian fencing schemes for reducing contributions from eroding channel banks to the siltation of Salmonid spawning gravels across the south west UK. J Environ Manag 91:1341–1349

Collins A, Zhang Y, MChesney D, Waling D, Haley S, Smith P (2012) Sediment source tracing in a lowland agricultural catchment in southern England using a modified procedure combining statistical analysis and numerical modelling. Sci Total Environ 414:301–317

Collins A, Zhang Y, Duethmann D, Walling D, Black K (2013) Using a novel tracing-tracking framework to source fine-grained sediment loss to watercourses at sub-catchment scale. Hydrol Process 27:959–974

Croft DJ, Pye K (2004) Colour theory and the evaluation of an instrumental method of measurement using geological samples for forensic applications. In: Pye K, Croft DJ (Eds) Forensic geosciences: principles, techniques, and applications, vol 232. Geological Society, London, Special Publications, pp 49–62

Damiani D, Giorgetti G (2008) Provenance of glacialmarine sediments under the McMurdo/Ross Ice Shelf (Windless Bight, Antarctica): heavy minerals and geochemical data. Palaeogeogr Palaeocl 260:262–283

Dearing JA, Hu YQ, Doody P, James PA, Brauer A (2001) Preliminary reconstruction of sediment-source linkages for the past 6000 yrs at the Petit Lac d'Annecy, France, based on mineral magnetic data. J Paleolimnol 25:245–258

Deasy C, Quinton J (2010) Use of rare earth oxides as tracers to identify sediment source areas for agricultural hillslopes. Solid Earth 1:111–118

de Boer DH, Crosby G (1995) Evaluating the potential of SEM/EDS analysis for fingerprinting suspended sediment derived from 2 contrasting topsoils. Catena 24:243–258

de Boer DH, Stone M, Levesque LMJ (2000) Fractal dimensions of individual flocs and floc populations in streams. Hydrol Process 14:653–667

Decou A, Mamani M, von Eynatten H, Wörner G (2009) Geochemical and thermochronological signals in tertiary to recent sediments from the Western Andes (1519 S): proxies for sediment provenance and Andean uplift. Geophys Res Abstr 11

D'Haen K, Verstraeten G, Degryse P (2012) Fingerprinting historic fluvial sediment fluxes. Prog Phys Geogr 36:154–186

Douglas GB, Gray CM, Hart BT, Beckett R (1995) A strontium isotopic investigation of the origin of suspended particulate matter (SPM) in the Murray-Darling river system, Australia. Geochim Cosmochim Acta 59:3799–3815

Douglas G, Palmer M, Caitcheon G (2003) The provenance of sediments in Moreton Bay, Australia: a synthesis of major, trace element and Sr-Nd-Pb isotopic geochemistry, modeling and landscape analysis. Hydrobiologia 494:145–152

Dudley RJ, Smalldon KW (1978) The objective comparison of the particle size distribution in soils with particular reference to the sand fraction. Med Sci Law 18:278–281

Eberl DD (2004) Quantitative mineralogy of the Yukon River system: changes with reach and season, and determining sediment provenance. Am Miner 89:1784–1794

Evrard O, Nemery J, Gratiot N, Duvert C, Ayrault S, Lefevre I et al (2010) Sediment dynamics during the rainy season in tropical highland catchments of central Mexico using fallout radionuclides. Geomorphology 124:42–54

Evrard O, Poulenard J, Némery J, Ayrault S, Gratiot N, Duvert C, Prat C, Lefévre I, Bonté P, Esteves M (2013) Tracing sediment sources in a tropical highland catchment of central Mexico by using conventional and alternative fingerprinting methods. Hydrol Process 27:911–922

Foster I, Lees J, Owens P, Walling D (1998) Mineral magnetic characterization of sediment sources from an analysis of lake and floodplain sediments in the catchments of the Old Mill reservoir and Slapton Ley, South Devon, UK. Earth Surf Proc Land 23:685–703

Fox JF, Papanicolaou AN (2008a) Application of the spatial distribution of nitrogen stable isotopes for sediment tracing at the watershed scale. J Hydrol 358:46–55

Fox J, Papanicolaou A (2008b) An un-mixing model to study watershed erosion processes. Adv Water Resour 31:96–108

Franks SW, Rowan JS (2000) Multi-parameter fingerprinting of sediment sources: uncertainty estimation and tracer selection. In: Bentley LR, Brebbia CA, Gray WG, Pinder GF, Sykes JF (eds) Computational methods in water resources. Balkema, Rotterdam, pp 1067–1074

Gaspar L, Navas A, Walling D, Machín J, Gómez Arozamena J (2013) Using ^{137}Csand^{210}Pb$_{ex}$ to assess soil redistribution on slopes at different temporal scales. Catena 102:46–54

Gellis A, Walling D (2011) Sediment source fingerprinting (tracing) and sediment budgets as tools in targeting river and watershed restoration programs. Geophys Monogr Ser 194:263–291

Gellis A, Hupp C, Pavich M, Landwehr J, Banks W, Hubbard B, Langland M, Ritchie J, Reuter J (2009) Sources, transport, and storage of sediment at selected sites in the Chesapeake Bay Watershed. US Geological Survey

Gingele FX, De Deckker P (2005) Clay mineral, geochemical and Sr-Nd isotopic fingerprinting of sediments in the Murray-Darling fluvial system, southeast Australia. Aust J Earth Sci 52:965–974

Giosan L, Flood RD, Grutzner J, Mudie P (2002) Paleoceanographic significance of sediment color on western North Atlantic drifts: II. Late Pliocene-Pleistocene sedimentation. Mar Geol 189:43–61

Gleason JD, Finney SC, Peralta SH, Gehrels GE, Marsaglia KM (2007) Zircon and whole-rock Nd-Pb isotopic provenance of middle and upper Ordovician siliciclastic rocks, Argentine Precordillera. Sedimentology 54:107–136

Golosov V, Belyaev V, Markelov M (2013) Application of Chernobyl-derived ^{137}Cs fallout for sediment redistribution studies: lessons from European Russia. Hydrol Process 27:781–794

Gotte T, Richter DK (2006) Cathodoluminescence characterization of quartz particles in mature arenites. Sedimentology 53:1347–1359

Gotze J, Plotze M, Habermann D (2001) Origin, spectral characteristics and practical applications of the cathodolumimescence (CL) of quartz a review. Miner Petrol 71:225–250

Grimes CB, John BE, Kelermen PB, Mazdab FK, Wooden JL, Cheadle MJ et al (2007) Trace element chemistry of zircons from oceanic crust: a method for distinguishing detrital zircon provenance. Geology 35:643–646

Grimshaw DL, Lewin J (1980) Source identification for suspended sediments. J Hydrol 47:151–162

Guzmán G, Barrn V, Gómez JA (2010) Evaluation of magnetic iron oxides as sediment tracers in water erosion experiments. Catena 82:126–133

Guzmán G, Quinton JN, Nearing MA, Mabit L, Gómez JA (2013) Sediment tracers in water erosion studies: current approaches and challenges. J Soil Sediment 13:816–833

Hallsworth CR, Chisholm JI (2008) Provenance of late Carboniferous sandstones in the Pennine Basin (UK) from combined heavy mineral, garnet geochemistry and palaeocurrent studies. Sediment Geol 203:196–212

Hancock GJ, Revill AT (2013) Erosion source discrimination in a rural Australian catchment using compound-specific isotope analysis (CSIA). Hydrol Process 27:923–923

Hardy F, Bariteau L, Lorrain S, Theriault I, Gagnon G, Messier D et al (2010) Geochemical tracing and spatial evolution of the sediment bed load of the Romaine River, Quebec, Canada. Catena 81:66–76

Hasholt B (1988) On identification of sources of suspended sediment transport in small basins with special references to particulate phosphorus. In: Proceedings of the symposium on sediment budgets, Porto Alegre, Brazil, pp 11–15

Hatfield RG, Maher BA (2009) Fingerprinting upland sediment sources: particle size-specific magnetic linkages between soils, lake sediments and suspended sediments. Earth Surf Proc Land 34:1359–1373

He Q, Owens P (1995) Determination of suspended sediment provenance using caesium-137, unsupported lead-210 and radium-226: a numerical mixing model approach. Sediment and water quality in river catchments, pp 207–227

Hillier S (2001) Particulate composition and origin of suspended sediment in the R. Don, Aberdeen-shire, UK. Sci Total Environ 265:281–293

Horwitz A (1991) A primer on sediment-trace element chemistry, 2nd edn. Lewis, Chelsea

Horowitz AJ, Stephens VC, Elrick KA, Smith JA (2012) Annual fluxes of sediment-associated trace/major element geochemistry of Lake Coeur d'Alene, Idaho, USA. Part II: subsurface sediments. Hydrol Process 9:35–54

Jenkins P, Duck R, Rowan J, Walden J (2002) Fingerprinting of bed sediment in the Tay Estuary, Scotland: an environmental magnetism approach. Hydrol Earth Syst Sci 6:1007–1016

Johnnson M (1993) The system controlling the composition of clastic sediments. In: Processes controlling the composition of clastic sediments. Geological Society of America Special Papers vol 284, pp 1–19

Juracek K, Ziegler A (2009) Estimation of sediment sources using selected chemical tracers in the Perry lake basin, Kansas, USA. Int J Sediment Res 24:108–125

Kemp P, Sear D, Collins A, Naden P, Jones I (2011) The impacts of fine sediment on riverine fish. Hydrol Process 25:1800–1821

Kimoto A, Nearing MA, Shipitalo MJ, Polyakov VO (2006) Multi-year tracking of sediment sources in a small agricultural watershed using rare earth elements. Earth Surf Proc Land 31:1763–1774

Kirkland C, Pease V, Whitehouse M, Ineson J (2009) Provenance record from Mesoproterozoic-Cambrian sediments of Peary Land, north Greenland: implications for the ice-covered Greenland Shield and Laurentian palaeogeography. Precambrian Res 170:43–60

Klages M, Hsieh Y (1975) Suspended solids carried by the Gallatin River of southwestern Montana: II. Using mineralogy for inferring sources. J Environ Qual 4:68–73

Knighton D (1998) Fluvial forms and processes: a new prospective. Arnold, London

Knox J (1987) Historical valley floor sedimentation in the upper Mississippi Valley. Ann Assoc Am Geogr 77:224–244

Knox J (1989) Rates of floodplain overbank vertical accretion. In: Hagedor J (ed) Floodplain Evolution, Abstract International Geomorphology Floodplain Symposium, Gottingen

Krein A, Petticrew E, Udelhoven T (2003) The use of fine sediment fractal dimensions and colour to determine sediment sources in a small watershed. Catena 53:165–179

Kurashige Y, Fusejima Y (1997) Source identification of suspended sediment from grain-size distributions: I. Application of nonparametric statistical tests. Catena 31:39–52

Lee S, Kim J, Yang D, Kim J (2008) Rare earth element geochemistry and Nd isotope composition of stream sediments, south Han River drainage basin, Korea. Quatern Int 176–177:121–134

Lewin J, Wolfenden P (1978) The assessment of sediment sources: a field experiment. Earth Surf Proc 3:171–178

Mabit L, Benmansour M, Walling D (2008) Comparative advantages and limitations of the fallout radionuclides ^{137}Cs, ^{210}Pb$_{ex}$ and ^{7}Be for assessing soil erosion and sedimentation. J Environ Radioactiv 99:1799–1807

Macklin M (1985) Flood-plain sedimentation in the upper Axe Valley, Mendip, England. Transactions of the Institute of British Geographers pp 235–244

Madhavaraju J, García J, Hussain S, Mohan S et al (2009) Microtextures on quartz grains in the beach sediments of Puerto Peñasco and Bahia Kino, Gulf of California, Sonora, Mexico. Rev Mex Cienc Geol 26:367–379

Maher BA, Watkins SJ, Brunskill G, Alexander J, Fielding CR (2009) Sediment provenance in a tropical fluvial and marine context by magnetic 'fingerprinting' of transportable sand fractions. Sedimentology 56:841–861

Mahler BJ, Bennett PC, Zimmerman M (1998) Lanthanide-labeled clay: a new method for tracing sediment transport in Karst. Ground Water 36:835–843

Martínez-Carreras N, Udelhoven T, Krein A, Gallart F, Iffly J, Ziebel J, Hoffmann L, Pfister L, Walling D (2010) The use of sediment colour measured by diffuse reflectance spectrometry to determine sediment sources: application to the Attert River catchment (Luxembourg). J Hydrol 382:49–63

Massoudieh A, Gellis A, Banks W, Wieczorek M (2013) Suspended sediment source apportionment in Chesapeake Bay watershed using Bayesian chemical mass balance receptor modeling. Hydrol Process 27:3363–3374

Miller JR, Orbock Miller SM (2007) Contaminated rivers: a geomorphological-geochemical approach to site assessment and remediation. Springer, Berlin

Miller J, Lord M, Yurkovich S, Mackin G, Kolenbrander L (2005) Historical trends in sedimentation rates and sediment provenance, Fairfield Lake, western North Carolina. JAWRA 41:1053–1075

Miller J, Mackin G, Lechler P, Lord M, Lorentz S (2013) Influence of basin connectivity on sediment source, transport, and storage within the Mkabela Basin, South Africa. Hydrol Earth Syst Sci 17:761–781

Morton A (1991) Geochemical studies of detrital heavy minerals and their application to provenance research. Geol Soc Lond Spec Publ 57:31–45

Morton AC, Hallsworth CR (1994) Identifying provenance-specific features of detrital heavy mineral assemblages in sandstones. Sediment Geol 90:241–256

Morton A, Fanning M, Milner P (2008) Provenance characteristics of Scandinavian basement terrains: constraints from detrital zircon ages in modern river sediments. Sediment Geol 210:61–85

Motha J, Wallbrink P, Hairsine P, Grayson R (2003) Determining the sources of suspended sediment in a forested catchment in southeastern Australia. Water Resour Res 39:1056–1069

Motha J, Wallbrink P, Hairsine P, Grayson R (2004) Unsealed roads as suspended sediment sources in an agricultural catchment in south-eastern Australia. J Hydrol 286:1–18

Mukundan R, Radcliff DE, Ritchie JC, Risse LM, McKinley RA (2010) Sediment fingerprinting to determine the source of suspended sediment in a southern Piedmont stream. J Environ Qual 39:1328–1337

Mukundan R, Walling D, Gellis A, Slattery M, Radcliffe D (2012) Sediment source fingerprinting: transforming from a research tool to a management tool. JAWRA 48:1241–1257

Nagle G, Fahey T, Ritchie J, Woodbury P (2007) Variations in sediment sources and yields in the Finger Lakes and Catskills regions of New York. Hydrol Process 21:828–838

Oldfield F (2007) Sources of fine-grained magnetic minerals in sediments: a problem revisited. Holocene 17:1265–1271

Oldfield F, Appleby PG, Van der Post KD (1999) Problems of core correlation, sediment source ascription and yield estimation in Ponsonby Tarn, West Cumbria, UK. Earth Surf Proc Land 24:975–992

Oszczypko N, Salata D (2005) Provenance analyses of the Late Cretaceous-Palaeocene deposits of the Magura Basin (Polish Western Carpathians) evidence from a study of the heavy minerals. Acta Geol Pol 55:237–267

Owens P, Walling D, Leeks G (1999) Deposition and storage of fine-grained sediment within the main channel system of the River Tweed, Scotland. Earth Surf Proc Land 24:1061–1076

Parsons A, Foster I (2011) What can we learn about soil erosion from the use of [137]Cs? Earth Sci Rev 108:101–113

Passmore D, Macklin M (1994) Provenance of fine-grained alluvium and late Holocene land-use change in the Tyne basin, northern England. Geomorphology 9:127–142

Peart M, Walling D (1986) Fingerprinting sediment source: the example of a drainage basin in Devon, UK. Drainage basin sediment delivery proceedings, Albuquerque 159:41–55

Phillips D, Gregg J (2003) Source partitioning using stable isotopes: coping with too many sources. Oecologia 136:261–269

Phillips J, Russell M, Walling D (2000) Time-integrated sampling of fluvial suspended sediment: a simple methodology for small catchments. Hydrol Process 14:2589–2602

Pirrie D, Butcher AR, Power MR, Gottlieb P, Miller GL (2004) Rapid quantitative mineral and phase analysis using automated scanning electron microscopy (QuemSCAN): potential applications in forensic geosciences. In: Pye K, Croft DJ (Eds) Forensic geosciences: principles, techniques, and applications, vol 232. Geological Society, London, Special Publications, pp 123–136

Pittam N, Foster I, Mighall T (2009) An integrated lake-catchment approach for determining sediment source changes at Aqualate Mere, Central England. J Paleolimnol 42:215–232

Polyakov VO, Nearing MA (2004) Rare earth element oxides for tracing sediment movement. Catena 55:255–276

Polyakov VO, Kimoto A, Nearing MA, Nichols MH (2009) Tracing sediment movement on a semiarid watershed using rare earth elements. Soil Sci Soc Am J 73:1559–1565

Porto P, Walling D, Callegari G (2005) Investigating sediment sources within a small catchment in southern Italy. In: Walling D, Horowitz A (eds) Sediment bugets I. IAHS Press no 291. Wallingford, UK, pp 113–122

Pye K (2004) Forensic examination of rocks, sediments, soils, and dust using scanning electron microscopy and X-ray chemical microanalysis. In: Pye K, Croft DJ (eds) Forensic geosciences: principles, techniques, and applications, vol 232. Geological Society, London, Special Publications, pp 103–122

Reynolds PH, Pe-Piper G, Piper DJW, Grist AM (2009) Single-grain detrital-muscovite ages from Lower Cretaceous sandstones, Scotian basin, and their implications for provenance. Bull Can Petrol Geol 57:63–80

Ritchie J, Ritchie C (2008) Bibliography of publications of ^{137}Cs studies related to erosion and sediment deposition. Medmrežje: http://www.ars.usda.gov/Main/docs.htm

Rollinson H (1993) Using geochemical data: evaluation, presentation, interpretation. Longman, Harlow

Rowan J, Goodwill P, Franks S (2000) Uncertainty estimation in fingerprinting suspended sediment sources. In: Foster IDL (ed) Tracers in geomorpholog pp 279–290

Rowan JS, Black S, Franks SW (2012) Sediment fingerprinting as an environmental forensics tool explaining cyanobacteria blooms in lakes. Appl Geogr 32:832–843

Russell M, Walling D, Hodgkinson R (2001) Suspended sediment sources in two small lowland agricultural catchments in the UK. J Hydrol 252:1–24

Singh P (2009) Major, trace and REE geochemistry of the Ganga river sediments: influence of provenance and sedimentary processes. Chem Geol 266:242–255

Slattery MC, Walden J, Burt TP (2000) Use of mineral magnetic measurements to fingerprinting suspended sediment sources: results from a linear mixing model. In: Foster IDL (ed) Tracers in geomorphology. Wiley, Chichester

Small I, Rowan J, Franks S (2002) Quantitative sediment fingerprinting using a Bayesian uncertainty estimation framework. Int Assoc Hydrol Sci Publ 276:443–450

Small IF, Rowan JS, Franks SW, Wyatt A, Duck RW (2004) Bayesian sediment fingerprinting provides a robust tool for environmental forensic geoscience application. In: Pye K, Croft DJ (eds) Forensic geoscience: principles, techniques and applications, vol 232. Geological Society, London, Special Publications, pp 207–213

Stuut J, Prins M, Schneider R, Weltje G, Jansen J, Postma G (2002) A 300-kyr record of aridity and wind strength in southwestern Africa: inferences from grain-size distributions of sediments on Walvis Ridge, SE Atlantic. Mar Geol 180:221–233

Syvitski J, Vörösmarty C, Kettner A, Green P (2005) Impact of humans on the flux of terrestrial sediment to the global coastal ocean. Science 308:376–380

Takeda A, Kimura K, Yamasaki S (2004) Analysis of 57 elements in Japanese soils, with special reference to soil group and agricultural use. Geoderma 119:291–307

Taylor A, Blake W, Couldrick L, Keith-Roach M (2012) Sorption behaviour of beryllium-7 and implications for its use as a sediment tracer. Geoderma 187–188:16–23

USEPA (1999) Protocol for developing Sediment TMDLs. Office of Water (4503F), United States Environmental Protection Agency, Washington DC, EPA 841-B-004, p 132

Veevers J, Saeed A (2007) Central Antarctic provenance of Permian sandstones in Dronning Maud Land and the Karoo Basin: integration of U-Pb and T-DM ages and host-rock affinity from detrital zircons. Sediment Geol 202:653–676

Vologina E, Kashik S, Sturm M, Vorob'eva S, Lomonosova T, Kalashnikova I, Khramtsova T, Toshchakov S (2007) Results of research into Holocene sediments of the South and Central basins of Lake Baikal (BDP-97 and short cores). Russ Geol Geophys 48:312–322

Walden J, Slattery M, Burt T (1997) Use of mineral magnetic measurements to fingerprint suspended sediment sources: approaches and techniques for data analysis. J Hydrol 202:353–372

Wallbrink P, Murray A (1993) Use of fallout radionuclides as indicators of erosion processes. Hydrol Process 7:297–304

Wallbrink P, Roddy B, Olley J (2002) A tracer budget quantifying soil redistribution on hillslopes after forest harvesting. Catena 47:179–201

Walling D (2013) Beryllium-7: the Cinderella of fallout radionuclide sediment tracers? Hydrol Process 27:830–844

Walling DE, Collins AL (2005) Sediment budgets 1. In: Walling DE, Horowitz AJ (eds) International Association of Hydrological Science Publications No 291 Wallingford UK pp 123–133

Walling D, He Q (1999) Using fallout lead-210 measurements to estimate soil erosion in cultivated land. Soil Sci Soc Am J 63:1404–1412

Walling D, Peart M (1979) Suspended sediment sources identified by magnetic measurements. Nature 281:110–113

Walling DE, Woodward JC (1992) Use of radiometric fingerprints to derive information on suspended sediment sources. In: Bogen J, Walling DE, Day T (eds) Erosion and sediment transport monitoring programmes in river basins. IAHS Publication No 210 IAHS Press, Wallingford, pp 153–164

Walling D, Woodward J (1995) Tracing sources of suspended sediment in river basins: a case study of the River Culm, Devon, UK. Mar Freshw Res 46:327–336

Walling DE, Woodward JC, Nicholas AP (1993) A multi-parameter approach to fingerprinting suspended sediment sources. In: Peters NE, Hoehn E, Leibundgut CH, Tase N, Walling DE (eds) Tracers in hydrology. IAHS Publication No 215 IAHS Press, Wallingford, pp 329–337

Walling D, Owens P, Leeks G (1999) Fingerprinting suspended sediment sources in the catchment of the River Ouse, Yorkshire, UK. Hydrol Process 13:955–975

Walling D, Collins A, McMellin G (2003a) A reconnaissance survey of the source of interstitial fine sediment recovered from Salmonid spawning gravels in England and Wales. Hydrobiologia 497:91–108

Walling D, Owens P, Foster I, Lees J (2003b) Changes in the fine sediment dynamics of the Ouse and Tweed basins in the UK over the last 100–150 years. Hydrol Process 17:3245–3269

Walling D, Schuller P, Zhang Y, Iroumé A (2009) Extending the timescale for using beryllium 7 measurements to document soil redistribution by erosion. Water Resour Res 45. doi:10.1029/2008WR007143

Walling DE, Golosov V, Olley J (2013) Introduction to the special issue 'Tracer Applications in Sediment Research'. Hydrol Process 27:775–780

Wang YJ, Zhao GC, Xia XP, Zhang YH, Fan WM, Li C et al (2009) Early Mesozoic unroofing pattern of the detrital zircon geochronology and Si-in-white mica analysis of synorogenic sediments in the Jianghan Basin. Chem Geol 266:231–241

Weltje G (2004) A quantitative approach to capturing the compositional variability of modern sands. Sediment Geol 171:59–77

Weltje G (2012) Quantitative models of sediment generation and provenance: state of the art and future developments. Sediment Geol 280:4–20

Weltje GJ, Prins MA (2003) Muddled or mixed? Inferring palaeoclimate from size distributions of deep-sea clastics. Sediment Geol 162:39–62

Weltje GJ, Prins MA (2007) Genetically meaningful decomposition of grain-size distributions. Sediment Geol 202:409–424

Weltje G, von Eynatten H (2004) Quantitative provenance analysis of sediments: review and outlook. Sediment Geol 171:1–11

Wilkinson S, Prosser I, Rustomji P, Read A (2009) Modelling and testing spatially distributed sediment budgets to relate erosion processes to sediment yields. Environ Model Softw 24:489–501

Wilkinson S, Hancock G, Bartley R, Hawdon A, Keen R (2013) Using sediment tracing to assess processes and spatial patterns of erosion in grazed rangelands, Burdekin River basin, Australia. Agr Ecosyst Environ 180:90–102

Wolman M, Miller J (1960) Magnitude and frequency of forces in geomorphic processes. J Geol pp 54–74

Wood PA (1978) Fine-sediment mineralogy of source rocks and suspended sediment, Rother catchment, West Sussex. Earth Surf Proc Land 3:255–263

Wood P, Armitage P (1997) Biological effects of fine sediment in the lotic environment. Environ Manag 21:203–217

Wude Y, Zhaoqian W, Guoping S, Guangwei D (2008) Quantitative determination of red-soil erosion by an Eu tracer method. Soil Till Res 101:52–56

Xu ZK, Lim D, Choi J, Yang SY, Jung H (2009) Rare earth elements in bottom sediments of major rivers around the Yellow Sea: implications for sediment provenance. Geo-Mar Lett 29:291–300

Yang SY, Jiang SY, Ling HF, Xia XP, Sun M, Wang DJ (2007) Sr-Nd isotopic compositions of the Changjiang sediments: implications for tracing sediment sources. Sci China Ser D 50:1556–1565

Yang XP, Zhang F, Fu XD, Wang XM (2008) Oxygen isotopic compositions of quartz in the sand seas and sandy lands of northern China and their implications for understanding the provenances of aeolian sands. Geomorphology 102:278–285

Yu L, Oldfield F (1989) A multivariate mixing model for identifying sediment source from magnetic measurements. Quatern Res 32:168–181

Yu LZ, Oldfield F (1993) Quantitative sediment source ascription using magnetic measurements in a reservoir-catchment system near Nijar, Se Spain. Earth Surf Proc Land 18:441–454

Zhang W, Xing Y, Yu L, Feng H, Lu M (2008) Distinguishing sediments from the Yangtze and Yellow Rivers, China: a mineral magnetic approach. Holocene 18:1139–1145

Zhang Y, Collins A, Horowitz A (2012) A preliminary assessment of the spatial sources of contemporary suspended sediment in the Ohio River basin, United States, using water quality data from the NASQAN programme in a source tracing procedure. Hydrol Process 26:326–334

Chapter 3
Fallout Radionuclides

Abstract A number of short-lived radioactive isotopes of both natural and anthropogenic origins which are (or were) atmospherically deposited over the landscape have been extensively utilized as sediment tracers in riverine environments. The three most extensively utilized isotopes, which are often referred to as fallout radionuclides (or FRNs), include in decreasing order of application, ^{137}Cs, $^{210}Pb_{ex}$, and ^{7}Be. Herein we examine the primary ways in which these three isotopes have been applied to gain insights into the riverine sediment system. More specifically, we explore the strengths and weaknesses of using FRNs in combination with mixing models to determine sediment provenance at the catchment scale, particularly with regards to determining whether the sediment was derived by means of sheet, rill, gully, or bank erosion. The nuclide inventory approach is also examined for its ability to characterize other components of the sediment system at much smaller spatial scales, including the redistribution of sediment on hillslopes and between landscape units. Our discussion concludes by examining the ability of ^{7}Be to document dynamic processes operating along the channel bed by determining sediment residence times, scour and fill depths, particle filtration, and sediment travel distances.

Keywords Fallout radionuclides · Hillslope sediment redistribution · Sediment residence time

3.1 Introduction

In the previous chapter we examined the use of geochemical data to quantitatively determine the relative contributions of sediment from defined sources at the catchment scale using inverse modeling techniques. A number of short-lived radioactive isotopes also have a long history of being used as tracers to address various components of the sediment system.

The field of isotope geochemistry was founded shortly after 1886 when Henri Becquerel discovered 'radioactivity' (a term put forth by Marie Curie) (Baskaran 2011). Since then about 300 stable and over 1,200 unstable isotopes have been identified (Hoefs 2010), and their use has become a central component of the analysis of Earth and environmental systems. Of particular importance has been the use of isotopes to constrain the age and rates of geological processes and as tracers to assess a

© The Author(s) 2015
J.R. Miller et al., *Application of Geochemical Tracers to Fluvial Sediment*,
SpringerBriefs in Earth Sciences, DOI 10.1007/978-3-319-13221-1_3

wide range of topics in almost every geological setting (Porcillie and Baskaran 2011), including the reconstruction of paleoclimates and sea level change, the paleochemistry of the oceans, the rates of weathering and erosion, the analysis of magmatic and tectonic processes, and the evolution of Earth and planetary systems, to mention just a few.

With regards to sediment and sediment-associated contaminants in river systems, the fallout radionuclides (FRNs) (e.g., ^{137}Cs, ^{210}Pb$_{ex}$, and ^{7}Be) have been most extensively utilized, often in combination with inverse mixing models, for source ascription purposes. The use of FRNs, however, is not restricted to the determination of sediment provenance by means of geochemical fingerprinting. Rather, they have been extensively used to characterize other important components of the sediment system. In the following sections we examine the general characteristics and spatial distribution of selected FRNs. We then turn our attention to the use of FRNs as a geochemical tracer of sediment within a catchment.

3.2 Lead-210, Cesium-137 and Beryllium-7: General Characteristics

The potential to use ^{137}Cs to indirectly determine average rates of soil loss was recognized as early as the 1960s (Menzel 1960; Rogowski and Tamura 1965). Since then, FRNs, particularly ^{137}Cs, ^{210}Pb$_{ex}$ and ^{7}Be, have been extensively investigated and used as geochemical tracers. Cesium-137 has been the most widely studied and applied FRN; in fact, a bibliography produced by Ritchie and Ritchie (2008) identified approximately 4,500 papers related to the transport and fate of Cs in the environment, a significant number of which pertain to the use of ^{137}Cs as a tool for analyzing erosional and depositional processes.

Cesium-137 is a 'man-made' or artificial radionuclide with a half-life of 30.2 years that primarily owes its presence in the environment to nuclear weapons testing and, to a much lesser extent, accidents at nuclear power plants (e.g., Chernobyl in 1986) (Ritchie and McHenry 1990; Mabit et al. 2008) (Table 3.1). Although the first weapons tests were carried out in 1945, the released ^{137}Cs was primarily retained within the region. In 1952, however, the testing of thermonuclear weapons injected ^{137}Cs into the stratosphere where it was distributed globally before being redeposited over the landscape (Perkins and Thomas 1980). Total fallout of ^{137}Cs was significantly (about an order of magnitude) greater in the northern hemisphere than in the southern hemisphere, reflecting the distribution of testing activities, and reached detectable concentrations in many areas in 1954 (Ritchie and McHenry 1990). Measurable concentrations in the southern hemisphere did not occur until about 1958. Concentrations in both hemispheres began to decrease following the 1963 Test Ban Treaty. By the mid-1970s, fallout had declined below detectable levels in the southern hemisphere, whereas in the northern hemisphere it could no longer be measured after 1983/1984 (Ritchie and McHenry 1990). The atmospheric deposition of ^{137}Cs occurs by both wet and dry processes, but the majority is associated with rainfall.

Table 3.1 General characteristics of ^{137}Cs, ^{210}Pb$_{ex}$, and ^{7}Be (adapted from Mabit et al. 2008 and Walling et al. 2011)

Characteristic	^{137}Ce	^{210}Pb$_{ex}$	^{7}Be
Origin	Man-made; nuclear weapons testing; power plant releases	Natural geogenic	Natural cosmogenic
Half-life	30.2 years	22.3 years	53.3 days
Timeframe of application	Since 1954; Markers in 1963/1964; 1986	$< \sim$100–150 years	Days $<$ 6 months
Temporal fallout pattern	Began in 1954; peaked in 1963/1964 in N. and S. hemispheres, respectively; ceased in late 1970s except for Chernobyl accidental release	Continuous with limited inter-annual variations	Continuous; high input variability with increases associated with precipitation events
Estimated process rate calculations	Annual averaged rates	Annual averaged rates	Events or short-periods of rainfall; daily to monthly averages for some residence time studies
Global pattern of fallout	High in N. hemisphere; low in S. hemisphere	Unknown	Unknown
Depth distribution			
• Undisturbed terrain	Peak at or immediately below surface; exponential decrease with depth; limited to about the upper 20 cm	Peak at surface; exponential decrease with depth; limited to about 30 cm	Peak at surface; exponential decrease with depth; limited to upper 10 cm or less
• Cultivated field	Uniform through plough zone	Uniform through plough zone	Exponential decrease from surface; non-detectable in recently cultivated land
Scale of area studied	Plot to watershed	Plot to watershed	Plot to short reach scale
Sample collection	Simple	Simple	Requires fine incremental sampling that can be difficult
Required analytical equipment	Normal HPGe γ detector	Broad energy range Normal HPGe γ detector	Normal HPGe γ detector

Thus, the rate of fallout and its spatial distribution is closely linked to the amount and intensity of precipitation (Longmore 1982; Basher and Matthews 1993).

Cesium-137 released during nuclear power plant accidents, such as Chernobyl, enters the troposphere where its deposition is primarily controlled by atmospheric circulation and precipitation patterns during and immediately following the event. Its distribution in soils, then, is much more heterogeneous and local than that associated with thermonuclear weapons testing. The distribution of ^{137}Cs from the Chernobyl incident, for example, affected large parts of Russia as well as other areas in the surrounding regions (Fig. 3.1) (Golosov et al. 1999, 2013), but cannot generally be detected in soils located further from the site.

The cycling of ^{137}Cs in the near surface environment plays a key role in its use as a geochemical tracer, and has been reviewed by Ritchie and McHenry (1990) and, more recently, by Parsons and Foster (2011) and Mabit et al. (2013) (Fig. 3.2). In general, atmospheric ^{137}Cs is deposited directly on soils (and other geological materials), upon water bodies, and on vegetation. A portion of the ^{137}Cs deposited on vegetation is adsorbed where it may be washed off during subsequent precipitation events and incorporated into the surrounding soils. The remainder of the Cs deposited on vegetation is absorbed by the plants where it is transferred to the soil by litterfall, or by incorporating Cs into the soil as plants die and decay. Since the Cs in the overlying vegetation eventually reaches the ground surface, the amount of ^{137}Cs within an undisturbed soil profile that has been subjected to no erosion or deposition

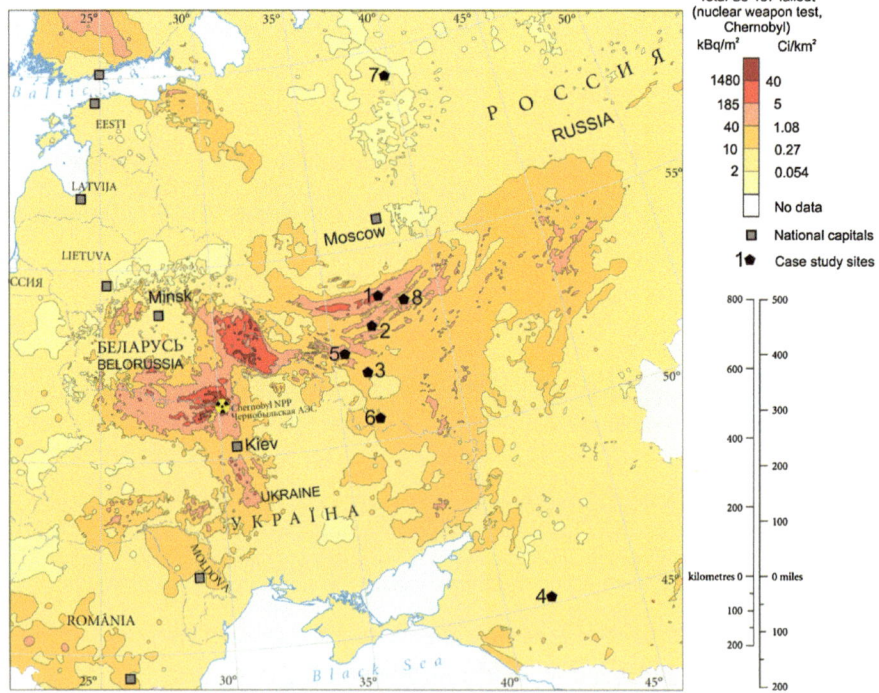

Fig. 3.1 Map showing ^{137}Cs (total, recalculated to 10 May 1986) in Eastern Europe and the locations of the case study sites examined in Golosov et al. (2013), including: (1) Lokna River basin; (2) Zusha River basin; (3) Vorobzha River basin; (4) Kalaus River basin; (5) Chern River basin; (6) Severskiy Donets River basin; (7) Toshnya River basin; and (8) Turdei River basin (from Golosov et al. 2013)

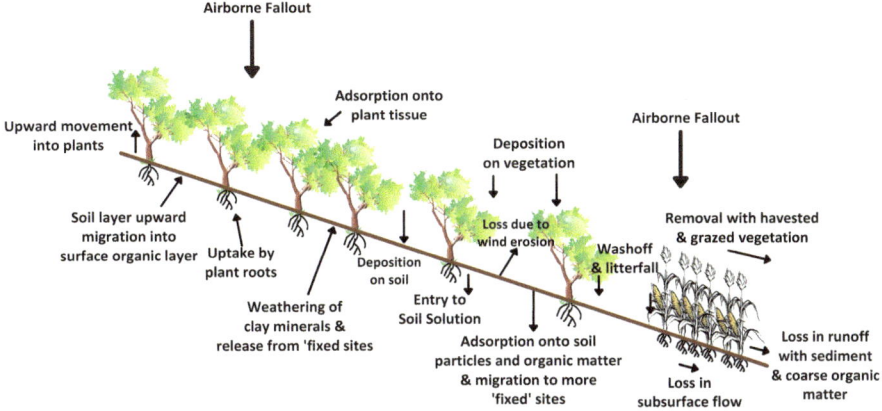

Fig. 3.2 Pathways of ^{137}Cs dispersal through the soil-plant system (from Parsons and Foster 2011)

since the time of ^{137}Cs fallout will closely approximate the total atmospheric ^{137}Cs flux to the Earth's surface.

A key assumption in the use of ^{137}Cs as a geochemical tracer is that it is strongly bound to fine-grained particles in the upper 30 cm of the soil profile (Ritchie et al. 1974; Wise 1980; Walling and Quine 1992). This assumption has been supported by a number of investigations (e.g., Lomenick and Tamura 1965) and is widely accepted in the fingerprinting literature. However, study of the potential health effects of ^{137}Cs following the Chernobyl incident has begun to question the assertion, noting that Cs may be released from clay minerals by weathering processes and/or desorbed and taken up by plant roots under some physiochemical conditions (Parsons and Foster 2011) (Fig. 3.2). In either case, ^{137}Cs typically exhibits an exponential decrease in concentrations with depth in undisturbed areas, although slightly lower values may be observed immediately below the ground surface (within the upper 5 cm) (Fig. 3.3a). The lower and/or near constant values at the surface reflect the cessation of new ^{137}Cs fallout since the 1970s/1980s, dilution associated with the addition of plant litter, and the influence of bioturbation processes that mix the upper soil layers (Walling and Woodward 1995; Mabit et al. 2008). In agricultural areas, the ^{137}Cs profile is altered by the mixing of surface and deeper soil materials, a process that creates semi-uniform concentrations throughout the plough layer. Below the plough layer, ^{137}Cs activity tends to decrease exponentially (Fig. 3.3b).

In contrast to ^{137}Cs, ^{210}Pb is a natural geogenic radioisotope produced as part of the ^{238}U decay series. The half-life of ^{210}Pb is 22.26 years, and the ^{210}Pb activity is such that it can continue to be measured in soils for a period of about 4–5 half-lives, or about 100 years (Table 3.1). Its immediate parent along the decay chain is ^{222}Rn, a gas formed from the decay of ^{226}Ra. Most of the produced ^{222}Rn remains in the soil and decays to ^{210}Pb. Since this ^{210}Pb in the soil is created within the profile, and is in equilibrium with ^{226}Ra, it is referred to as supported ^{210}Pb. A small portion of the ^{222}Rn, however, diffuses into the atmosphere where it decays to ^{210}Pb before being

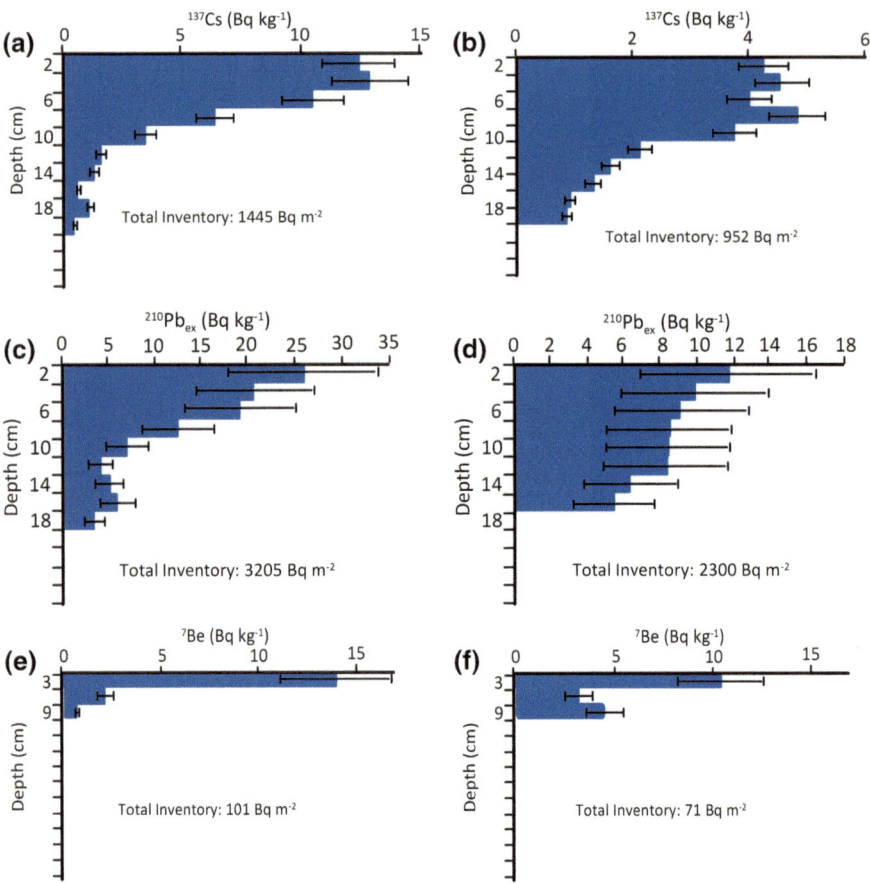

Fig. 3.3 Variations in fallout radionuclides with depth in soils from Morocco. **a** [137]Cs in an undisturbed soil; **b** [137]Cs in a ploughed agricultural soil; **c** [210]Pb$_{ex}$ in an undisturbed soil; **d** [210]Pb$_{ex}$ in a ploughed agricultural soil; **e** [7]Be in an undisturbed soil; **f** [7]Be in a ploughed agricultural soil (adapted from Mabit et al. 2008)

returned to the Earth's surface within a few days (Fig. 3.4). This atmospherically deposited ^{210}Pb is referred to as unsupported or excess [210]Pb$_{ex}$ to distinguish it from the ^{210}Pb created by, and in equilibrium with, [226]Ra in soil. The amount of supported ^{210}Pb in the soil can be determined on the basis of the [226]Ra activity and, thus, removed from the total to determine the unsupported [210]Pb$_{ex}$ in the soil derived from atmospheric deposition.

Like ^{137}Cs, an undisturbed soil profile that has been subjected to no erosion or deposition is thought to reflect the atmospheric [210]Pb$_{ex}$ flux to the Earth's surface. Unlike ^{137}Cs, however, the [210]Pb$_{ex}$ is essentially constant through time (although minor variations in the flux of [210]Pb$_{ex}$ to the Earth's surface have been noted; Preiss et al. 1996). Thus, while the inventory of ^{137}Cs in an undisturbed soil that has been

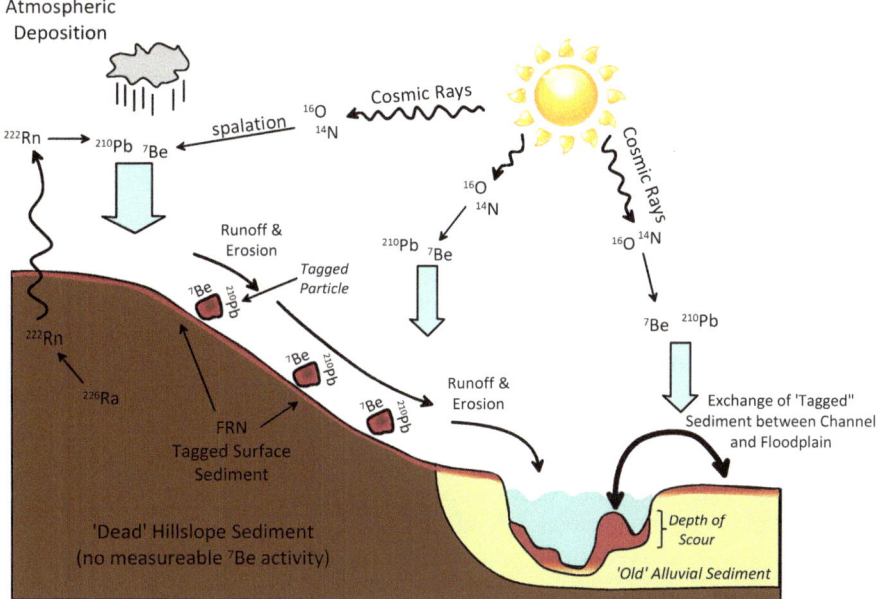

Fig. 3.4 Schematic diagram of the production, deposition, and use of ^{210}Pb and ^{7}Be as sediment tracers in drainage basins

subjected to no erosion will decline through time as a result of the cessation of fallout and radioactive decay after the 1970s/1980s, the continuous flux of ^{210}Pb$_{ex}$ to the soil creates a steady-state condition in which atmospheric deposition is balanced by radioactive decay (Zhang et al. 2006). Spatially, atmospheric depositional rates exhibit considerable variability; on a global scale fluxes are typically lower where air masses have traversed oceanic bodies and have had little chance of accumulating ^{222}Rn from soils and other geological materials. Higher depositional rates are associated with circulating air masses that have traveled over continental areas and have had an opportunity to accumulate ^{222}Rn. Locally, depositional rates are strongly influenced by geographical patterns in the amount and intensity of rainfall.

As was the case of ^{137}Cs, ^{210}Pb$_{ex}$ is assumed to be strongly and irreversibly attached to soil particles upon deposition from the atmosphere (Dickin 1997). Thus, ^{210}Pb$_{ex}$ exhibits a vertical profile very similar to that of ^{137}Cs where the highest concentrations are located at the ground surface and decrease exponentially with depth (Fig. 3.3c). A zone of uniform or lower concentration immediately below the ground surface does not occur as a result of the continuous and current flux of ^{210}Pb to the soil. In agricultural areas, ^{210}Pb$_{ex}$ concentrations exhibit semi-uniform values through the plough zone as a result of mixing of the upper horizons (Fig. 3.3d).

Beryllium-7 is a naturally produced, cosmogenic radionuclide created by cosmic ray bombardment of atmospheric N and O (Fig. 3.4). Thus, the production of ^{7}Be is linked to the cosmic ray flux, a parameter that varies with solar activity and changes in the Earths magnetic field as well as altitude and latitude (Kaste et al. 2002; Walling 2013). On a global scale, production is about three to five times higher at the poles than the equator. Regionally, Wallbrink and Murray (1994) found that more than 90 % of ^{7}Be is deposited as wet deposition and, thus, annual fallout varies primarily in response to annual precipitation amounts. For a given precipitation amount, annual fluxes are lower for lower latitudes (Walling 2013).

Although it is assumed that ^{7}Be is strongly bound to particulate matter upon reaching the soil, few studies have actually examined this postulate. Taylor et al. (2012), however, found by means of laboratory experiments that ^{7}Be was rapidly sorbed to unsaturated soil particles and remained attached to soils under commonly observed field conditions. ^{7}Be differs, however, from ^{137}Cs and ^{210}Pb in two important ways (Table 3.1). First, it has a short half-life of 53.3 days. Second, in part due to its short half-life, it is concentrated in the upper most soil layers, generally within the upper 1 cm (Fig. 3.3e) within undisturbed soils. Within this shallow zone, it typically exhibits an exponential downward decrease in concentrations. In agricultural areas, this exponential downward trend will occur in soils that have not be recently ploughed because of its continuous flux to the soil surface, but will be non-detectable in recently tilled soils as a result of mixing and dilution with deeper soil particles devoid of ^{7}Be (Walling 2013).

3.3 Applications

3.3.1 Use as a Geochemical Tracer to Determine Sediment Provenance

As was the case for other types of geochemical fingerprinting parameters, FRNs have been applied to determine the provenance of suspended, channel, and reservoir materials (Nagle et al. 2007; Huisman and Karthikeyan 2012; Kwan Kim et al. 2013). However, their vertical distribution in surface soils makes them particularly well suited to estimate the relative contribution of sediment derived from surface and subsurface sources. More specifically, in undisturbed areas, maximum concentrations of ^{137}Cs, ^{210}Pb$_{ex}$ and ^{7}Be occur at, or in the case of ^{137}Cs, immediately below the ground surface. Concentrations subsequently decrease exponentially typically reaching non-detectable levels within about 10, 20, and 30 cm of the surface for ^{7}Be, ^{210}Pb$_{ex}$, and ^{137}Cs, respectively. The overwhelming majority of the total inventory, then, is confined to the upper few centimeters of the Earth's surface. As a result, surface materials eroded by sheet or shallow-rill processes exhibit high concentrations of the FRNs, whereas subsurface materials eroded from gullies, channel or bank deposits, dirt roads, or any other barren surface where the upper few centimeters of sediment have been stripped away will often contain little or no detectable FRNs (Wallbrink and Murray 1993; Olley et al. 2013).

It is not uncommon for investigators to use ^{137}Cs, $^{210}Pb_{ex}$ and ^{7}Be to estimate the relative contribution of sediment from both spatially defined areas (e.g., forests, grazing areas, cropped land) and source types (e.g., channel and/or gully sediment) (see, for example, Nagle et al. 2007; Olley et al. 2013). In this case, the emphasis generally is on determining the contribution of sediment from the channel or channel banks in comparison to surface sources. In contrast to the multivariate geochemical fingerprinting approach discussed in the previous chapter, investigators often use a single FRN (e.g., Nagle et al. 2007; Kwan Kim et al. 2013). There are, however, benefits, as shown later, to utilizing a combination of ^{137}Cs, $^{210}Pb_{ex}$ and ^{7}Be (Olley et al. 2013; Walling 2013). When a single parameter is used as a fingerprint, a relatively simply mixing model has historically been applied for source ascription purposes (Zhang and Zhang 1995; Wallbrink et al. 1998; Zhang et al. 1997; Wallbrink and Murray 1996; Wallbrink et al. 1998; Brigham et al. 2001). Mathematically, it takes the form of (Wallbrink and Murray 1996):

$$C_s = ((P_r P_b)/(P_s - P_b)) \times 100 \qquad (3.1)$$

where C_s is the percent relative contribution from a surface source, P_s is the value of the FRN for the surface source, P_b is the value of the FRN in the subsurface source, and P_r is the value of the FRN in the river sediment (sediment mixture). The initial studies used mean FRN values to characterize the source materials. However, FRN concentrations in surface sediments tend to vary within the catchment as a function of atmospheric fallout and sediment transfers between sites as a result of erosional and depositional processes. These variations were initially dealt with by using a composite sampling scheme where multiple samples were collected and combined within a defined area to help eliminate field variance. While composite sampling is still common practice, more recent studies (e.g., Nagle et al. 2007) utilize a Monte Carlo approach similar to that described in the previous chapter to quantitatively account for model uncertainty. The FRNs for a specific source are characterized when using this approach by a statistical distribution of observed values within the materials. Values from these distributions are then selected and entered into the model to determine the relative contribution from surface sources. The process is repeated thousands of times, producing a statistical distribution of source contributions (C_s in Eq. 3.1). This generated distribution can subsequently be analyzed to determine the median predicted contribution (the 50th percentile) from the surface sources, while other quartiles can be used to assess the extent of variation in the predicted values.

It is normally assumed that FRN activities follow a normal distribution within the source and river sediments. Olley et al. (2013), however, found that within subtropical catchments of Queensland, Australia FRN data within the source and river sediments did not necessarily follow a normal distribution. In this case, only $^{210}Pb_{ex}$ was normally distributed. Thus, they used a procedure similar to that proposed by Caitcheon et al. (2012) to generate a probability distribution for ^{137}Cs that could be incorporated into a relatively simple mixing model to estimate sediment provenance. They found that channel deposits represented the dominant source of sediment to the river (as opposed to surface sources consisting of cropped land, forests, and grazing areas). The point to be made is that characterization of source end members may require the collection of more samples than originally thought, and it cannot be assumed a priori that FRN concentrations are normally distributed within the source and river sediments.

A commonly untested assumption is that ^{137}Cs and ^{210}Pb$_{ex}$ exhibit negligible concentrations below the first 20–30 cm of the floodplain surface. In most studies, FRN concentrations within the bank materials are determined on composite samples, consisting of sediments collected from the bank. While ^{137}Cs and ^{210}Pb$_{ex}$ concentrations within these samples may be low, the observed values may result from the mixing of upper bank sediments enriched in FRNs with lower bank sediments that exhibit concentrations below detection. Studies that have used ^{137}Cs and ^{210}Pb$_{ex}$ to date floodplain deposits illustrate, for example, that both radionuclides may extend to greater depths below the ground surface and occur at higher concentrations than found in upland soils. The higher concentrations result from the combined accumulation of radionuclides in the floodplain from atmospheric fallout as well as from the deposition of FRN-bearing sediments eroded from the surface of upland soils during overbank events (He and Walling 1996; Stokes and Walling 2003). The maximum concentrations of ^{137}Cs in the floodplain deposits will correspond to the deposition of sediment in 1963 in the northern hemisphere (1964 in the southern). Thus, the depth of maximum ^{137}Cs activity will vary as a function of the sedimentation rate on the floodplain since that time. The activity of ^{210}Pb$_{ex}$ will likely decrease exponentially below the floodplain surface, but because the surface layers may be episodically buried during floods, the trend may be chaotic and extend to greater depths. During bank erosion, these enriched surface deposits will add to the FRN activity measured within the river sediment, which depending on their vertical distribution and concentration within the floodplain and the bank height, may lead to an overestimation of sediment from upland sources.

While the above discussion focuses on ^{137}Cs and ^{210}Pb$_{ex}$, recent studies have shown that ^{7}Be can serve as an effective tracer. Nonetheless, Walling (2013) points out that there are several factors that complicate its use as a fingerprinting parameter. First, ^{7}Be has the potential to change rapidly through time in response to its fallout during storm events and its relatively rapid radioactive decay. Thus, the characterization of ^{7}Be in the source materials will need to be done at or very close to the time the river sediments are collected. This problem may pose considerable limitations on the sources of suspended sediments by inhibiting its application to materials that are no more than a few weeks old. Second, its concentration within the uppermost soil materials (generally $\sim$1 cm) and its rapid decrease below the surface means that the concentration of ^{7}Be within the eroded sediments will vary significantly as a function of the depth of erosion. While this is also true for ^{137}Cs and ^{210}Pb$_{ex}$, the variations are likely to be much more pronounced, thereby adding additional uncertainty to the discrimination of sediment sources by means of ^{7}Be (Walling 2013). Third, because a significant portion of the ^{7}Be may be intercepted by and stored on vegetation, substantial variations in ^{7}Be, even within a single land-use category, may occur as a result of variations in the vegetation cover. Finally, spatial variations in ^{7}Be may be initiated by the timing of human activities. Cultivation and mixing of the upper soil horizons, for example, may 'reset' the ^{7}Be inventory to non-detectable values. Thus, the ^{7}Be inventory of recently cultivated fields may differ significantly from fields that were cultivated some weeks in the past.

Although the use of a single FRN has proven to be effective at determining sediment provenance, the combined use of ^{137}Cs, ^{210}Pb$_{ex}$ and/or ^{7}Be can often lead to an improved ability to distinguish between sediment sources. Wilkinson et al. (2009), for example, effectively combined the use of ^{137}Cs and ^{210}Pb$_{ex}$ to distinguish between burnt and unburnt surface and subsurface sediments within a eucalypt-forested catchment near Sydney, Australia (Fig. 3.5a). Similarly, Walling (2013) illustrated how the combined use of ^{7}Be and ^{137}Cs could be used to more effectively distinguish between cultivated and uncultivated fields eroded by sheet, rill, and bank processes. In this latter case, ^{7}Be was able to discriminate between channel bank sediments, rill-eroded cultivated lands, eroded trackways, and the sheet erosion of pasture and cultivated soils. However, a distinction could not be made between sediments eroded by sheet erosion on pastures and cultivated lands without the additional use of ^{137}Cs (Fig. 3.5b).

3.3.2 Determination of Sediment Redistribution and Erosion Rates

3.3.2.1 Methods and Approach

One of the most widely used applications of ^{137}Cs as a geochemical tracer is to quantify the amount of erosion and deposition that has occurred at a specific location, that when combined with data from other sites can be used to assess the spatial redistribution of sediment across the landscape. The use of tracers other than ^{137}Cs have also been explored for this purpose during the past several decades, including ^{134}Cs (e.g., Syversen et al. 2001), ^{59}Fe (e.g., Wooldridge 1965), ^{239,240}Pu (Everett et al. 2008; Dong et al. 2010; Smith et al. 2012), ^{210}Pb$_{ex}$ (Wilkinson et al. 2009; Smith et al. 2012), and ^{7}Be (e.g. Blake et al. 1999; Walling et al. 1999; Schuller et al. 2006; Wilson et al. 2003; Kaste et al. 2011). The latter two (^{210}Pb$_{ex}$ and ^{7}Be) have received the most attention and will be focused upon here with ^{137}Cs.

The approach is similar for each of the above mentioned radionuclides. The total nuclide inventory for an undisturbed site that has experienced no erosion or deposition, called the *reference site*, is determined and compared to inventories measured at sites where the amount of soil erosion or deposition is in question. Since erosion or deposition has not occurred at the reference site, the inventory is assumed to be a function of the total atmospheric flux of the nuclide to the ground surface and its subsequent radioactive decay. In contrast, the inventory at a geomorphically disturbed site is a function of three primary factors: the atmospheric flux to the site, radioactive decay, and the loss or gain of a radionuclide associated with soil particles that are either eroded from or deposited at the site. Thus, negative or positive differences between the inventories of the reference and non-reference sites reflect

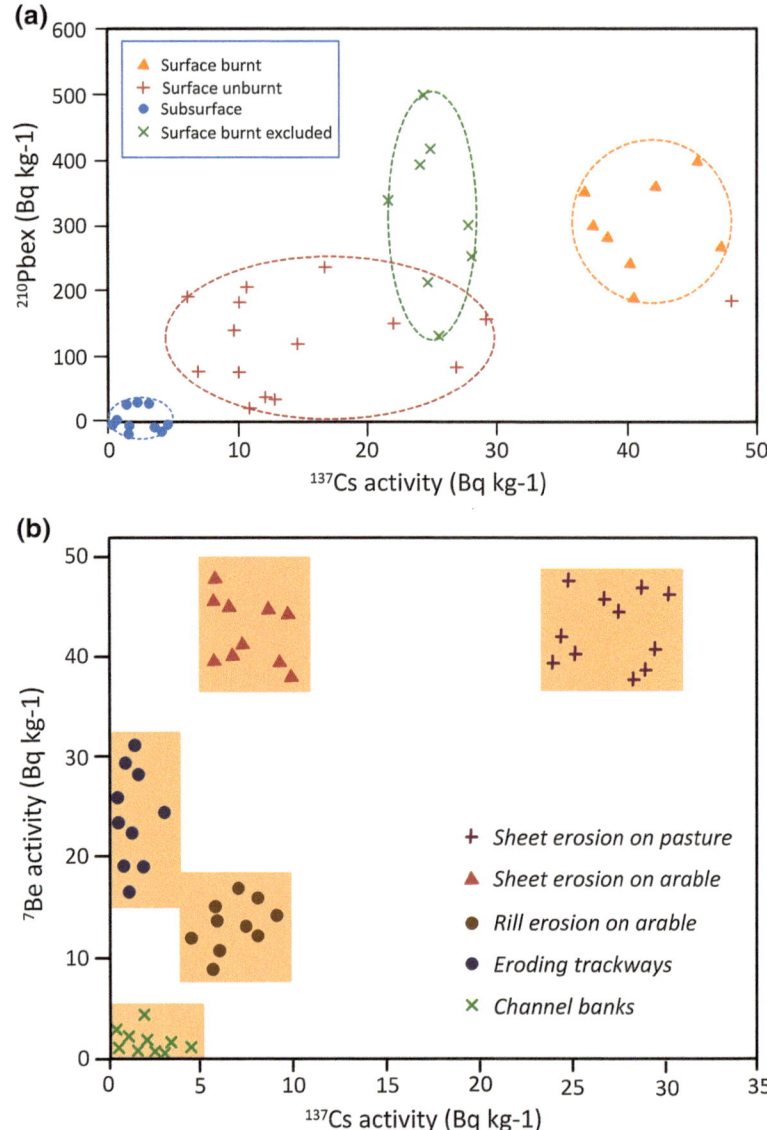

Fig. 3.5 **a** Bivariate plot illustrating the ability of ^{137}Cs and ^{210}Pb$_{ex}$ to discriminate between burnt and unburnt surface and subsurface sediments within a catchment of the Nattai River near Sydney, Australia. Radionuclide concentrations were measured on the <10 m particle-size fraction and were corrected for radioactive decay from May 2002 (figure from Wilkinson et al. 2009); **b** Illustration of the use of ^{7}Be and ^{137}Cs to discriminate between sources types within cultivated and uncultivated fields within a small catchment in Mid Devon, UK (from Walling 2013)

the amount of erosion or deposition that has occurred, respectively. By quantifying these differences, the magnitude of erosion or deposition since the onset of nuclide accumulation can be estimated.

Advantages and constraints on the use of this approach have been summarized by Mabit et al. (2008), Parsons and Foster (2011), Mabit et al. (2013), and, for ^{7}Be, Walling (2013). Three of the most commonly cited advantages are that (1) the method does not involve the use of long-term monitoring equipment, but can quantify erosional or depositional magnitudes using data/samples collected in a single trip, (2) the estimates of soil loss are based on specific sampling locations that when combined with data from other sampling locations can provide information on the spatial variations in erosional and depositional magnitudes across the landscape as well as for specific landscape elements, and (3) the approach provides retrospective information on past erosional and depositional processes, something that contemporary monitoring programs are incapable of doing. In addition, the use of more than one radionuclide, each characterized by a different half-life, allows the magnitude of erosion or deposition to be determined for a range of timeframes (Table 3.1).

Early studies, beginning in the 1960s and 1970s were primarily confined to the plot scale, but more recent studies have refined the approach and applied it to larger scales. Gaspar et al. (2013), for example, combined ^{137}Cs and ^{210}Pb$_{ex}$ data to determine medium to long-term soil redistribution rates at the hillslope scale within the Pre-Pyrenean Mountains of northeastern Spain. The upper part of the studied, covered by forest vegetation and characterized by an average slope of 24 %, was found to be relatively stable, exhibiting minor amounts of erosion and deposition (Fig. 3.6). The midslope, characterized by a wide range of land-use/land-cover types (including patches of forest and terraced fields) was dominated by minor erosion over the long-term ($\sim$100 years) as determined by ^{210}Pb$_{ex}$; ^{137}Cs suggested that over the medium term, erosion predominated, although local deposition also occurred. The highest rates of erosion as assessed by ^{137}Cs were associated with actively cultivated areas, whereas abandoned fields exhibited lower rates of erosion. In fact, some abandoned fields, along with the forested, midslope areas, were generally stable. Deposition dominated the bottom of the slope characterized by cultivated fields but a much reduced slope (15 %). Significant erosion was locally noted at a site below a pathway, suggesting that cultivation has affected soil mobilization along the bottom of the slope as well. Although minor differences exist, there was general agreement in the predominant processes occurring on the slope over the two timeframes provided by ^{137}Cs and ^{210}Pb; in both cases, the data reflect local land-use and slope characteristics. The detailed magnitudes of erosion and deposition were found, however, to vary in a complex way across the hillslope through time.

A recent modification to the approach is to establish a sediment budget for a hillslope or catchment on the basis of the FRN inventories. Sediment budgets are often required for the development of effective watershed management plans as they provide important insights into the primary sources of sediment that should be targeted for remediation, the transfer of sediment between landscape elements within the catchment, and the amount of sediment exported from the basin mouth (Dietrich and Dunne 1978; Kelsey et al. 1981; Reid et al. 1981; Trimble 1983).

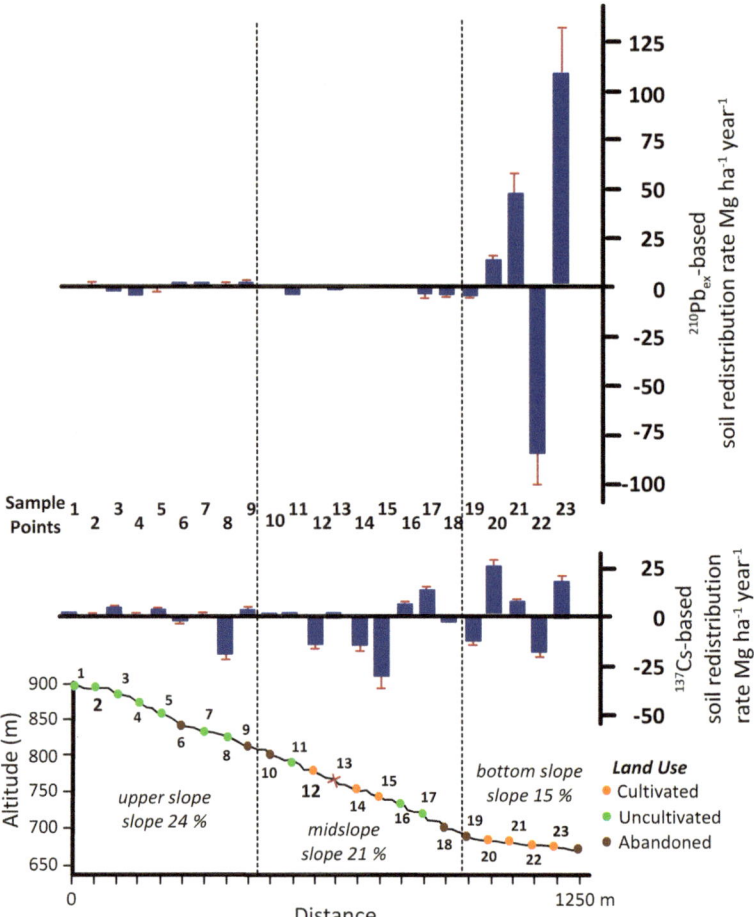

Fig. 3.6 Variations in estimated soil redistribution rates calculated using ^{137}Cs and ^{210}Pb for a slope in the Pre-Pyrenean Mountains of northeast Spain. Local profile shows sampling points, average hillslope gradient and land-use at sampled site (from Gaspar et al. 2013)

However, sediments budgets are difficult to develop using traditional monitoring techniques because while they can effectively document sediment transport rates along the drainage network, they do not explicitly determine the transfer of sediment from one landscape unit to another (Blake et al. 2009). The FRN budgeting approach overcomes this problem.

Wallbrink et al. (2002), for example, demonstrated that a sediment budget could be constructed for a hillslope on the basis of ^{137}Cs inventories to assess areas of soil erosion and deposition, the rates of sediment transfer between different landform units, and the overall loss of sediment from the system. In this case, the data were needed to assess the effectiveness of sediment management practices associated with forest harvesting. In general, their sediment budget, and those that followed, was based on three measured terms. First, the initial inventories, or the amount of the tracer present in each of the landscape elements of interest prior to disturbance must be known. These initial inventories are usually assessed using the radionuclide activities found at a reference site that is extrapolated to the whole of the catchment. Second, the area covered by the defined landscape elements of interest must be determined. Third, the current inventory present within the landscape elements must be measured, a value that represents the amount of tracer left following disturbance (Wallbrink et al. 2002). The product of the tracer inventories and area measured for a landscape element is then assumed to be the total amount of each tracer in that element, and the value used in constructing the sediment budget. Results from the study are shown in Fig. 3.7, and led Wallbrink et al. (2002) to conclude that the practice of dispersing flow and sediment from the highly compacted snig tracks to less compacted and spatially larger general harvest areas and filter strips was effective at retaining surface soil and sediment mobilized from a harvested area. It is important to note that when the sediment budgeting approach is being used to assess the effects of a specific disturbance (forest harvesting in the case of Wallbrink et al. 2002), then the amount of erosion and/or deposition that has occurred between the initial timing of radionuclide fallout and the beginning of the disturbance regime must be known.

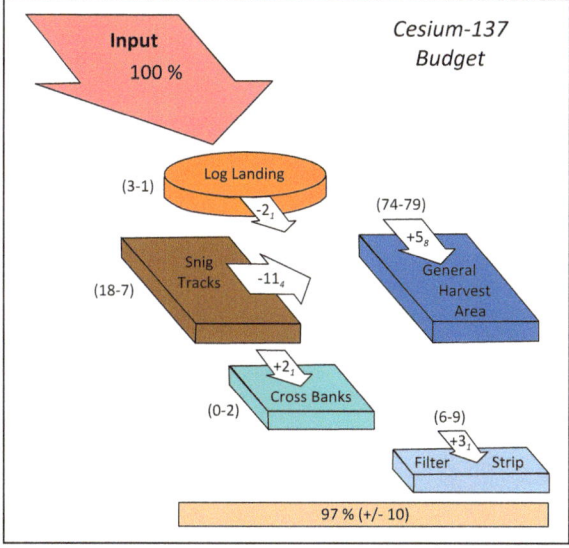

Fig. 3.7 Tracer budget based on ^{137}Cs constructed by Wallbrink et al. (2002) about 6 years after forest harvesting within a small catchment of the Bondi State Forest, New South Wales, Australia. The relative amount of sediment contained within each landscape unit is shown in parentheses. The relative percent of the total sediment input that was transported out of or deposited within the landscape units are shown by values within the *arrows*. Cumulative errors associated with measurement errors and element surface area derivations are shown by subscripts (from Wallbrink et al. 2002)

This quantity is often difficult to determine, and, as described in more detail below, can represent a significant constrain on the use of ^{7}Be as a tracer as a result of its relatively short half-life (Walling 2013).

Blake et al. (2009) applied the FRN budgeting approach to a small (89 ha), wildfire-affected catchment in SE Australia. The objective of the study was to assess the magnitude of the post-fire sediment and nutrient redistribution that had occurred. Actually, three budgets were constructed using ^{210}Pb$_{ex}$, ^{137}Cs, and ^{7}Be, representing average erosional and depositional processes over approximately the past 60 years, 50 years, and 3 months, respectively (Fig. 3.8) (Blake et al. 2009). All three tracers found that ridgetop and sideslope units were the primary sources of sediment to the drainage network. However, notable differences exist between the constructed budgets (Fig. 3.8). Blake et al. (2009) argued that these differences not only reflect differences in the timeframe under consideration, but the materials with which the isotopes were associated and the depth to which the radionuclides extend below the surface. ^{137}Cs was associated with subsurface mineral matter, whereas ^{210}Pb$_{ex}$ and ^{7}Be were primarily associated with ash, litter, and soil organic matter at or near the ground surface. Thus, ^{137}Cs tended to reflect the erosion of deeper, coarser grained mineral matter, while ^{210}Pb$_{ex}$ and ^{7}Be reflected the movement of litter, ash and burnt O-horizon materials. Estimated sediment export ratios for ^{210}Pb$_{ex}$ and ^{7}Be suggest that 96–99 % of the eroded organic material from the surface reached the stream channel (Fig. 3.8). In contrast, only 67 % of the deeper mineral matter was transported to the channel, the remainder presumably deposited on the slope and/or valley floor. The point to be made here is that the combined use of the three FRNs not only provided insights into the magnitude of the erosion that occurred over different timeframes and the primary sources of sediment to the channel network, but the nature of the sediment remobilization processes.

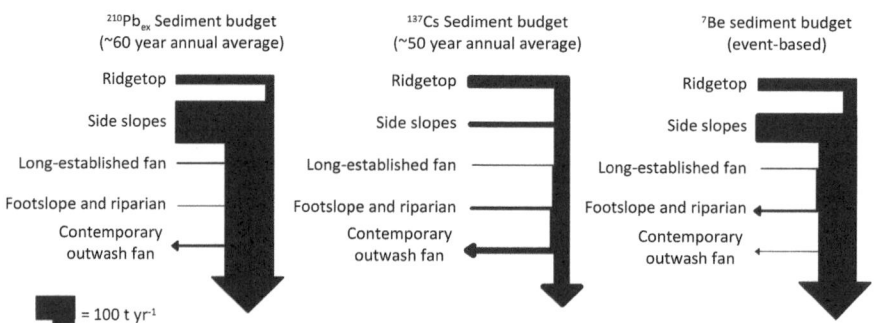

Fig. 3.8 Schematic of annual average sediment budgets derived from ^{210}Pb$_{ex}$, ^{137}Cs, and ^{7}Be tracer budgets for a small wildfire affected catchment in southeastern Australia. Thickness of *arrows* is approximately proportional to total mass of sediment transferred over the respective timeframes represented by the radionuclides. The budgeting approach allowed for the estimation of sediment transfers between landscape units as well as the amount of sediment exported from the catchment (from Blake et al. 2009)

3.3.2.2 Uncertainties

The assessment of erosional and depositional magnitudes using of the FRN inventory approach has been applied in a wide range of geologic, physiographic, and climatic settings, and is now widely accepted by many, if not most, in the scientific community. Acceptance of the method is based in part on the validation of its results using data provided by (1) traditional approaches such as erosion plots, erosion pins, and the monitoring of catchment sediment yields (e.g., Elliott et al. 1990; Porto et al. 2001, 2003a, b; Mabit et al. 2002; Wallbrink and Croke 2002; Schuller et al. 2006; Ceaglio et al. 2012), empirical erosion models (e.g., the USLE and RUSLE) (Di Stefano et al. 1999; Saç et al. 2008; Mabit et al. 2008), and other FRNs (e.g., $^{210}Pb_{ex}$, $^{239,240}Pu$) (Hoo et al. 2011; Ketterer et al. 2011; Matisoff and Whiting 2011). Moreover, trends in the generated results often fit what is intuitively expected on the basis of land-use and physiographic characteristics of the site. Nonetheless, some investigators have questioned the validity of the generated results. Parsons and Foster (2011), for example, concluded that no current rates of soil erosion that are based upon the use of this technique are reliable, and that ^{137}Cs cannot be used to provide reliable information about rates of soil erosion (see Mabit et al. 2013 for a rebuttal). Many of concerns raised by Parsons and Foster (2011) are linked to three of the fundamental assumptions upon which the FRN method (including ^{137}Cs) is based, including that: (1) ^{137}Cs was uniformly distributed over the landscape, and its distribution was not modified before it was bound to sedimentary particles, (2) ^{137}Cs is rapidly, strongly and irreversibly bound to sediment and/or organic materials, such that the observed differences in its inventories reflect the movement of soil particles, and (3) estimates of soil erosion or deposition can be accurately estimated from measurements of ^{137}Cs inventories using the available conversion models. These assumptions are explored in more detail below.

1. *Cesium-137 (and presumably other FRNs) are Initially Distributed Uniformly Across the Study Area*
   An important assumption inherent in the approach is that the FRNs are initially distributed uniformly across the study area. This allows for any spatial deviation in inventories from the reference site to be assigned to the redistribution of the FRNs after their deposition on and sorption to soil particles. Parsons and Foster (2011) argued that the assumption of an initially uniform distribution of ^{137}Cs (and, presumably other FRNs) is invalid. To support their contention, they argued that if the assumption was strictly observed the variability in FRNs between cores of the reference sites would be similar. However, in the case of ^{137}Cs, spatial variations in inventories of as much as 40 % have been recorded (Wallbrink et al. 1994; Sutherland 1994; Wallbrink and Murray 1996). Parsons and Foster (2011) note that part of the variability is related to the non-uniform atmospheric deposition of ^{137}Cs (and presumably $^{210}Pb_{ex}$, ^{7}Be, etc.) by means of wet deposition as a result of topographic and geographical patterns in precipitation amounts and intensities. In other words, variations in precipitation across the landscape are likely to result in the spatially non-uniform fallout of ^{137}Cs and other FRNs to the soil surface.

It is generally accepted, however, that over time differences in rainfall patterns between precipitation events will reduce or eliminate the variations in fallout (Walling and Quine 1992). This 'smoothing' effect will be most effective when deposition occurs over periods of years (e.g., for ^{137}Cs derived from bomb tests) in comparison to deposition over periods of months (e.g., ^{137}Cs from Chernobyl or ^{7}Be). The degree of heterogeneity is also likely to vary as a function of spatial scale, becoming more significant at larger spatial scales, where large differences in elevation exist, or where reference sites are located far from the study area. Other factors cited by Parsons and Foster (2011) that may lead to spatial variations in FRN content include (1) the indirect transfer of FRNs from the atmosphere to the soil as a result of the existing vegetation cover (Fig. 3.2) (Dörr and Münnich 1987; Wallbrink and Murray 1996), and (2) the redistribution of FRNs by flow processes prior to attaching to the soil particles (particularly after the infiltration capacity of the soil has been exceeded) (Lance et al. 1986; Foster et al. 1994).

2. *FRNs are rapidly, strongly, and irreversible bound to sediment and organic matter*
   An important assumption in the use of FRNs as tracers is that they are rapidly, strongly and irreversibly fixed to soil particles (Walling and He 1999). This assumption is required to attribute spatial variations in FRN inventories to the redistribution of soil particles by means of erosional and depositional processes. While Cs mobility in soils is a complex process controlled by multiple parameters (e.g., pH, organic matter content, particle mineralogy, CEC), the assumption has been widely accepted by the tracer community (e.g., Ritchie et al. 1974; Walling and Quine 1992; Nouira et al. 2003; Mabit et al. 2008). Parsons and Foster (2011), however, suggested using data collected primarily to assess the potential ecological and human health effects of ^{137}Cs from environmental incidents (e.g., Chernobyl) that ^{137}Cs is not as rapidly, strongly, or irreversibly bound to sediment or organic matter as originally assumed. Their argument is based on studies/data that suggest (1) the sorption of ^{137}Cs to clay minerals varies from one mineral type to another, and may be "a rather slow process which extends over many years" (Bunzl et al. 1995). Thus, at least some of the ^{137}Cs that reaches the soil may remain in the soil solution and be redistributed with the migrating fluids; (2) ^{137}Cs is partly associated with exchangeable sorption sites on particle surfaces, indicating that it is not permanently bound to the particles (Livens et al. 1996), and (3) the nature of the sorption sites to which ^{137}Cs is bound changes through time, initially being associated with more available planar sites and becoming less bioavailable as it becomes bound in the interlayer sites in a process referred to as radiocesium ageing (Staunton 1994). This ageing process may be counteracted by mineral weathering that releases the ^{137}Cs from the strongly bound sites (Wendling et al. 2005). Desorption from the soil particles may also be promoted by a decrease in Cs concentrations within the soil solution, a process that may occur in the vicinity of plant roots as Cs is taken up by plants (Fig. 3.2). The net effect of these processes is that cycling of ^{137}Cs in the near surface environment may lead to their redistribution by processes other than particle movement, or may result in the loss of FRNs from the system, producing either an overestimate of the initial inventory or an underestimation of the inventory within the disturbed

soils. However, Mabit et al. (2013) cogently argue that the ^{137}Cs (FRN) method is not based on the absolute amounts in the soil, but on a comparison between the reference and sample sites. Considering that the processes affecting the sample sites are also likely to affect the reference sites, lateral redistribution in soils is likely to be negligible in comparison to the movement of ^{137}Cs associated with physical or anthropogenic erosion processes.

3. *Estimates of soil erosion or deposition can be accurately estimated from measurements of the FRN inventories*
   Differences in FRN inventories between an undisturbed (reference) and disturbed site must be converted into an absolute quantity of eroded or deposited sediment. This conversion is no easy task given that the character of the soil (e.g., its organic matter content, grain size, and bulk density) will undoubtedly vary across the study area, and the FRN activity changes as a function of depth (Fig. 3.3). A wide range of models have been developed to tackle the problem over the past three decades (e.g., Elliott et al. 1990; Ritchie and McHenry 1990; Kachanoski 1987; Zhang et al. 1990; Fredericks and Perrens 1988; Walling and He 1999; Fornes et al. 2005; Soto and Navas 2008; Walling et al. 2002; Wallbrink and Murray 1996; Walling et al. 2011). These models differ in (1) their degree of complexity and sophistication, (2) whether they can be applied to cultivated or uncultivated areas, and (3) the FRN to which they are applicable.

   Selection of a conversion model is an important aspect of any FRN tracer study and is discussed in more detail in Walling et al. (2011), and Porto and Walling (2012). The selection process will necessarily need to consider the intended use of the resulting estimates of soil erosion and deposition, the land-use/land-cover at the study site, and the data that are available or can be obtained for use in the models. The point to be made here is that different models are likely to generate different results. Fornes et al. (2005), for example, found an approximately threefold difference between the results generated from two different conversion models when applied to ^{137}Cs data that had been collected in 1974 and again in 1998 at the National Soil Tilth Laboratory Deep Loess Research Station in Southwest Iowa. Part of the difference in modeling results in the Fornes et al. (2005) study is that one model incorporated variations in ^{137}Cs fallout through time, whereas the other did not. The comparison of model results show that in general the more complex models yield more reliable estimates of the amount of soil erosion and/or deposition that occurs for the time period of interest, but they require data that are often difficulty to obtain (Mabit et al. 2008). In addition, many of the more refined models are highly sensitive to variations in the input parameters requiring that the additional data be accurately collected.

3.3.2.3 Summary of the Limitation on FRN Inventory Derived Estimates

It is hard to argue that the assumptions required to estimate the magnitude of sediment redistribution on hillslopes or within a catchment using FRNs are strictly upheld. The question, however, is whether the departures from the methods inherent assumptions are so significant that they render the results meaningless. In answer to this question, we believe that FRNs can provide useful information about the magnitude (but perhaps not the absolute amount) of sediment redistribution upon a hillslope or within a small catchment for several reasons. First, as noted earlier, a number of studies have compared FRN generated results with data collected using more traditional approaches and have found them to be comparable. Second, the data consistently estimate magnitudes of sediment redistribution that systematically vary across the landscape in a manner that is consistent with our current understanding of how land-use/land-cover, geologic, and physiographic controls influence erosional and depositional processes. Third, estimates of sediment redistribution using ^{137}Cs and ^{210}Pb$_{ex}$ are often consistent. Given that the two elements were derived from different sources, were deposited at different times and rates, and possess different chemistries, particularly with regards to solution-particulate distribution coefficients, one could expect ^{137}Cs and ^{210}Pb$_{ex}$ to produce significantly different results if the underlying assumptions were not generally upheld. It should also be remembered that the more traditional approaches to measuring erosion rates (e.g., erosion pins and runoff plots) contain significant, and often unquantified uncertainties. Thus, at the present time it is possible that FRNs are as reliable as other utilized methods, and they represent one of the only methods to assess past magnitudes of upland erosion and deposition. We acknowledge, however, that the data should be interpreted and utilized with a large degree of caution, particularly with regards to estimates of the absolute magnitudes of erosion or deposition that has occurred.

3.3.2.4 Additional Constraints on the Use of ^{7}Be

Beryllium-7 has received considerably less attention than either ^{137}Cs or ^{210}Pb$_{ex}$ as a tracer in river basins (Matisoff and Whiting 2011). Nonetheless, it has the potential to serve as an important tracer because its relatively short half-life (53.3 days) allows for the documentation of sediment redistribution patterns and rates over periods of days to weeks, corresponding to individual flood events or short-periods of intense rainfall. Thus, ^{7}Be can serve as a complement to the use of ^{137}Cs or ^{210}Pb$_{ex}$ where the latter examines sediment sources and sinks over periods of decades (Walling 2013). The short half-life of ^{7}Be, however, places a number of additional restrictions on it use that significantly limits it application.

One of the most significant limitations of ^{7}Be is related to the portion of the total inventory that can be accumulated and stored on vegetation. In the case of ^{137}Cs or ^{210}Pb$_{ex}$, the total inventory represents an average flux to the soil over a period of decades. Fallout that is intercepted by vegetation is generally thought to reach the soil as it is washed off in subsequent events, or as plant materials fall to the ground surface and decay. Thus, only a small portion of the total inventory resides within the vegetation cover. In contrast, much of the ^{7}Be fallout may be intercepted by vegetation during an individual rainfall event and, because of its short half-life, remain there until it decays. It is possible, then, that much of the measurable ^{7}Be fallout never reaches the ground surface in vegetated areas. Several studies, for example, have shown that more than about 50 % of the fallout may be stored in the vegetation canopy in both grassland (Bettoli et al. 1995; Kaste et al. 2011) and forested (Kaste et al. 1999) areas. Moreover, variations in vegetation type, density, height, etc. are likely to lead to significant variations in the ^{7}Be flux to the underlying soil and the resulting total ^{7}Be inventory. The net effect is that the use of ^{7}Be as a tracer of sediment redistribution is generally limited to areas of bare ground where the influences of vegetation are absent.

The short half-life of ^{7}Be also leads to significant variations in total ^{7}Be inventories and penetration depths within the soil. Figure 3.9, for example, shows the variations in ^{7}Be inventories and relaxation depths for a reference site near Valdivia, Chile as reported by Walling et al. (2009). The depth-distribution curves correspond to samples collected over an 11 month period extending from January 24 to November 30, 2006. Soil inventories and relaxation depths are relatively low during the dry summer months (November through early June), but systematically increase during the wet season (late June to early November). The dynamic nature of the ^{7}Be inventory and depth distribution leads to two significant issues. First, estimates of soil erosion and deposition are typically determined for ^{7}Be using a profile distribution model developed for ^{137}Cs for uncultivated fields. The model requires the depth distribution of ^{7}Be to be known. Thus, the dynamic nature of the ^{7}Be inventory and depth distribution curves during an extended monitoring period will increase the uncertainty in the results, and limits the use of ^{7}Be to a period of a few weeks. Second, and

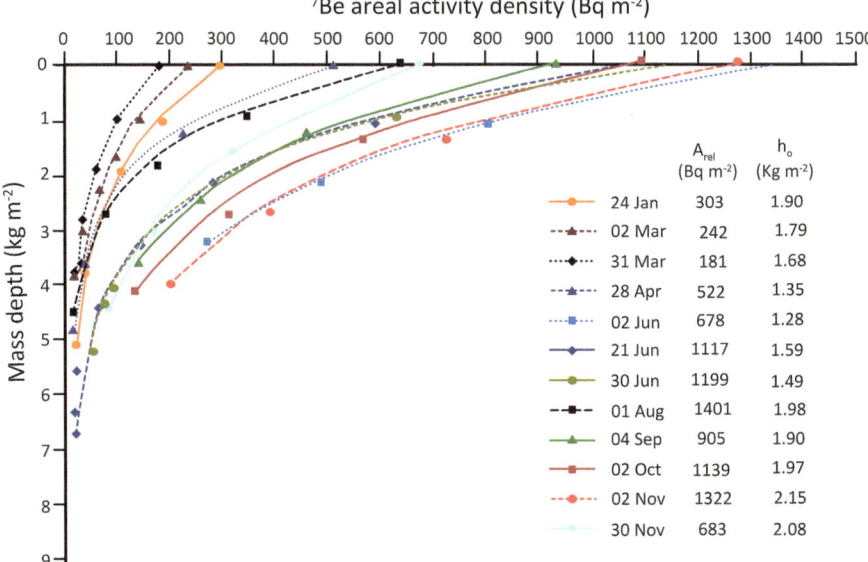

Fig. 3.9 Variations in ^{7}Be active as a function of depth below the ground surface for data collected on multiple dates at a reference site in Valdivia, Chile. Soil inventories (Aref) and relaxation depths (ho) are low during the dry summer months of November through early June, but increase during the wet season (late June to early November) (from Walling et al. 2009)

perhaps more importantly, attempts to determine the amount of sediment redistribution that has occurred during an individual event or short period of intense rainfall requires that the spatial distribution of ^{7}Be be uniform across the area (as is true for ^{137}Cs or ^{210}Pb$_{ex}$). However, the dynamic nature of ^{7}Be inventories, which can change in response to rainfall patterns or erosion/depositional processes during a single event, means that ^{7}Be is unlikely to exhibit a homogenous spatial distribution unless the inventories inherited from previous events have been eliminated by radioactive decay, or by means of cultivation and mixing of the soil profile (Walling 2013). In the absence of cultivation, Walling (2013) suggests that about 5 months may be required to eliminate past patterns in ^{7}Be inventories. In other words, it may be that ^{7}Be can only be applied as a tracer of sediment redistribution where significant rainfall events (i.e., those that redistribute sediment) have not occurred for the past 5 months, severely limiting its application.

3.4 Estimating Sediment Age, Residence Times, Transport Distances, and Other Contemporary Sedimentation Processes

Over the past 10–15 years there has been a growing interest in the erosional and depositional processes that operate along the channel bed, particularly with regards to the magnitude of channel bed scour and fill during flood events, the 'age' and residence time of the sediment that comprise the bed material, and distances and rates that particles move over a specified period of time. Interest in these processes stem from the fact that the exchange of particles with the channel bed as well as their travel times and distances during a given event or set of events strongly influences where the impact of sediment and its associated contaminants will be concentrated and for how long (Bonniwell et al. 1999). Long transport distances and short residence times, for example, indicate that the sediment and any associated pollutants will be passed rapidly downstream to receiving bodies (e.g., lakes, reservoirs, or estuaries), whereas shorter travel distances and longer residence times may be associated with more severe impacts on riverine habitats and biota. Quantifying these processes has proven to be difficult using direct methods, leading investigators to explore the use of FRNs to decipher such sedimentation processes. The majority of the work conducted to date focuses on contemporary processes at the reach- or river-corridor scale, thereby requiring the use of ^{7}Be (characterized by a short half-life).

The use of ^{7}Be (and other FRNs) to quantify channel bed sedimentation processes is based on the postulated existence of a relatively simply system. Exposed surfaces within the catchment (e.g., the surface of upland soils, floodplains, and emergent channel bars) are tagged by atmospherically deposited FRNs, including ^{7}Be (Fig. 3.4). Once the tagged sediments enter the flow, they can no longer receive FRNs from the atmosphere. FRN activity will then begin to decrease as a result of radioactive decay, and since the decay process is constant, the activity of the river sediments can serve as a measure of how recently the particles entered the channel.

Sediments that entered the flow recently will exhibit higher activities than those that have remained in the channel bed and have been inundated for an extended period of time (Kasprak et al. 2013). If the sediments remain in the channel bed, or are buried within alluvial deposits for a long-enough period of time, the FRN activity will decrease below detectable levels and the sediment is considered to be dead. In the case of ^{7}Be, radionuclide activity can be measured in sediments that have entered the flow within about 265 days, whereas it is about 112 years for ^{210}Pb characterized by a much longer half-life. Any re-exposure of the bed sediments, such as those associated with the surface of channel bars that are periodically inundated during floods, will be retagged with the FRNs. In theory, then, the atmospherically tagged sediments make an ideal tracer as the signature of the recently eroded and tagged particles differ significantly in radionuclide activity from the sediment already in the channel allowing the particles to be tracked through the drainage network (Bonniwell et al. 1999).

A number of studies have attempted to examine the time it takes for particles eroded from upland areas to move downstream through the drainage network. The estimation of sediment residence time, travel distances, and particle exchange rates with the channel bed are of particular concern. A central tenet of the approach is that the depositional fluxes to the ground surface are uniform over the catchment so that the eroded particles possess the same signature before they enter the flow no matter where they are derived. However, over short periods, depositional fluxes are highly variable, raising questions as to whether ^{7}Be can be used as an independent dating tool (Matisoff et al. 2005). Moreover, FRN activity can vary between grain size fractions. Thus, hydraulic sorting and other processes that partition sediments according to their size and composition into different parts of the channel may affect the measured ^{7}Be signature. To reduce the effects of varying atmospheric fluxes and sedimentological characteristics on the study results, it is common for the ^{7}Be activity to be normalized by ^{210}Pb activity. This normalization by ^{210}Pb is based on the logic that because ^{7}Be and ^{210}Pb are both derived from atmospheric fallout, are strongly bound to particulate matter, and exhibit a $+2$ valance state in aqueous solutions, they should exhibit a similar distribution in the environment. In fact, their activity should co-vary as a function of grain size and compositional differences. Thus, normalizing ^{7}Be activity by the activity of ^{210}Pb (^{7}Be/^{210}Pb) should both decrease the spatial variability observed for the individual isotopes (Matisoff et al. 2005), and partially correct for variations in activity associated with compositional differences (Bonniwell et al. 1999; Salant et al. 2007). The argument of a reduced spatial variability in the ^{7}Be/^{210}Pb ratio in comparison to that of the individual isotopes is consistent with the observations made by Baskaran et al. (1993) and Koch et al. (1996). In addition, because the half-life of ^{210}Pb is much larger than that of ^{7}Be, changes in the ratio will predominately reflect the radioactive decay of ^{7}Be and the age of the sediment.

It is important to note that the total ^{210}Pb activity, as opposed to ^{210}Pb$_{ex}$, is typically used for normalization. This stems from the fact that the amount of ^{210}Pb that is supported by ^{222}Rn cannot be determined in river systems where the loss of ^{226}Ra by advection in pore-waters migrating through the sediments inhibits a determination of the degree to which disequilibrium between ^{210}Pb and ^{226}Ra may occur (Salant et al. 2007). The use of total ^{210}Pb may be problematic because it includes ^{210}Pb from both lithogenic and atmospheric sources, the activity of which decreases with increasing grain size (Salant et al. 2007). However, the associated errors are not so significant that they are likely to invalidate the results.

Since the ^{7}Be/^{210}Pb ratio measured in a river's suspended load or channel bed sediment will reflect the time that has elapsed since the tagged sediment has entered the flow, the ratio can be compared to the signature of the sediment that enters the water to determine the sediments age, t. Age, in this case, is measured relative to the time the sediment first entered the water, and can be calculated using the following equation (Matisoff et al. 2005):

$$t = \frac{-1}{\lambda_{^7Be} - \lambda_{^{210}Pb}} \log\left(\frac{^7Be_{sample}}{^{210}Pb_{sample}}\right) + \frac{1}{\lambda_{^7Be} - \lambda_{^{210}Pb}} \log\left(\frac{^7Be_{source}}{^{210}Pb_{source}}\right) \quad (3.2)$$

where $^7Be_{source}$ and $^{210}Pb_{source}$ are the ^{7}Be and ^{210}Pb activities in the source material, $^7Be_{sample}$ and $^{210}Pb_{sample}$ are the activities in the sampled river sediment, and $\lambda_{^7Be}$ and $\lambda_{^{210}Pb}$ are the decay constants for ^{7}Be ($0.01300 d^{-1}$) and ^{210}Pb ($8.50999 \times 10^{-5} d^{-1}$), respectively. Spatial and temporal changes in the ^{7}Be/^{210}Pb in river sediments will also reflect dilution processes where 'new' sediment is exchanged for 'dead' sediment in the channel bed, or 'new' sediment is diluted with 'old' sediment eroded from the channel bed and banks. This relationship between ^{7}Be/^{210}Pb ratios and dilution processes allows for an estimation of the relative percent of 'new' sediment, S_{new} within the load or deposit to be calculated by (Matisoff et al. 2005):

$$S_{new} = 100 \cdot \frac{^7Be_{sample}/^{210}Pb_{sample}}{^7Be_{source}/^{210}Pb_{source}} \quad (3.3)$$

or

$$S_{new} = 100 \cdot e^{-\left(\lambda_{^7Be} - \lambda_{^{210}Pb}\right)t} \quad (3.4)$$

In addition to the estimation of sediment age and dilution effects, the analysis can be further extended to assess particle travel distances over specific time intervals by documenting downstream trends in either the ^{7}Be/^{210}Pb ratio or the percent of 'new' sediment within the suspended or bedload. For example, Bonniwell et al. (1999), following the approach provided by Cushing et al. (1993), found that ^{7}Be/^{210}Pb ratios collected on a given day varied semi-systematically along the Gold Fork River of Idaho. The data could be expressed by a non-linear function of the form:

$$F(x) = F_0 e^{-k_1 x} \quad (3.5)$$

where $F(x)$ is the measure ^{7}Be/^{210}Pb ratio in the sediment, F_0 is the ^{7}Be/^{210}Pb ratio of the source materials at the point of input, x is distance downstream, and k_1 is the rate of change along the river. Once F_0 and k_1 are determined from an equation fit to the measured data, the average transport distance can be calculated as $\frac{1}{k_1}$. The use of the equation is constrained, however, by two assumptions: (1) that the input of 'new' sediment, S_{new}, is associated with a point source, and (2) the system is in an equilibrium state in which there is no net storage or loss of sediment from the channel bed (Bonniwell et al. 1999). In the study of the Gold Fork River, Bonniwell et al. (1999) argued that both assumptions were reasonably met allowing the equation to be applied to their dataset. They found that particle transport distances varied from about 60 km near the peak of the snowmelt hydrograph to 12 km near base flow, a trend that matched expectations.

While the above methods have been used by a number of investigators (e.g., Bonniwell et al. 1999; Matisoff et al. 2005; Salant et al. 2007; Evrard et al. 2010) to determine sediment residence times, etc., a significant constraint on the approach is that changes in the $^7Be/^{210}Pb$ ratio along the channel at a given time reflect both the age of the sediment and dilution processes, a fact clearly recognized by the investigators. Matisoff et al. (2005), for example, argued that the dual control of sediment age and dilution on the ratios could be treated such that each represented end-member states. Thus, by analyzing them separately, the two analyses provide a unique perspective on the interpretation of the results. In reality, however, changes in the ratio are likely to be due to both processes, leading to a nearly infinite set of possible interpretations of the data. Determining which interpretation is correct will likely be met with difficulty and lace the results with a large amount of uncertainty.

Other key assumptions upon which the approach is based may also lead to large uncertainties in the results. For example, Walling (2013) points out that the assumed similarity between recently eroded and tagged sediment and that in fallout (rainfall) is unlikely to be met as the ratio will vary from year to year, seasonally, and from one storm to the next. Thus, the use of a constant value for the sediment entering the channel is unlikely to generate realistic results. Matisoff et al. (2005) suggest that this problem may be overcome in part by measuring the $^7Be/^{210}Pb$ ratio in precipitation. However, the $^7Be/^{210}Pb$ of eroded surface sediments will not only reflect recent fallout, but the inventories accumulated over decades in the case of ^{210}Pb. Thus, the $^7Be/^{210}Pb$ measured in precipitation will likely differ from that of the surface soils. The $^7Be/^{210}Pb$ ratio will also vary across the landscape as a result of erosional and depositional processes that deplete or increase the FRN activities, relative to that of an undisturbed site (as noted earlier). Sediment input into the river, then, may exhibit a range of $^7Be/^{210}Pb$ ratios, depending on where the sediment was derived. Finally, because the depth distribution of ^7Be and ^{210}Pb differ (Fig. 3.3), the $^7Be/^{210}Pb$ ratio may vary through time and space as a result of the depth to which erosion occurs (Bonniwell et al. 1999). Some of these problems in defining an effective source signature may be minimized, as Matisoff et al. (2005) suggest, by using the $^7Be/^{210}Pb$ ratio of the sediments eroded and suspended in runoff entering the channel, although the effectiveness of this approach has yet to be demonstrated.

In light of the above, it appears that estimates of sediment age and residence times are unlikely to provide meaningful results given the time and cost associated with the analysis unless the boundaries of the study are highly constrained. Fisher et al. (2010), for example, utilized ^7Be to quantitatively assess the storage times of sediment in bars associate with in-channel obstructions including boulder and large woody debris in the Ducktrap River of coastal Maine. Their study differs from earlier investigations in that (1) the predominant input of FRNs to the river was the wet deposition of the radionuclides to the surface of emergent bars located along the channel margins and not from external sources, and (2) storage times were based on the constant initial activity (CIA) decay model where the initial activities, A_0, were defined by a range of values equal to one standard deviation above and below the mean activities observed on adjacent thalweg and emergent bars at a site. The CIA model is expressed as:

$$t = \frac{\log \frac{A}{A_0}}{-\lambda} \tag{3.6}$$

where A is the observed activity of the sample and A_0 is the initial activity at the time of deposition. They found that sediment storage times were influenced by reach-scale variations in unit stream power and the frequency of large woody debris. Sediment storage associated with in-channel obstructions were generally longer (> 100 days) than along supply limited reaches which exhibited sediment storage times generally < 100 days. Use of the CIA model for these types of in-channel assessments require that a significant portion of the stored sediments are exposed to the atmospheric deposition of ^{7}Be and that the initial activities of the sediments can be estimated (Fisher et al. 2010). In addition, greater uncertainties may occur where dilution of bed or bar materials by dead sediment make it more difficult to define the initial ^{7}Be activities.

A different approach to the determination of sediment residence times utilizes a mass balancing method carried out at the catchment scale (Dominik et al. 1987; Le Cloarec et al. 2007; Evrard et al. 2010). Originally put forth by Dominik et al. (1987), the model subdivides sediment movement within a catchment into two components (referred to as boxes), a soil box in which sediment transfers are slow and residence times are long, and a river box in which sediment transfers occur rapidly and sediment residence times are short (Fig. 3.10). Data required for the model include an understanding of the atmospheric fluxes of ^{7}Be, ^{137}Cs, and ^{210}Pb, which represents radionuclide inputs to the catchment, and the ^{7}Be, ^{137}Cs, and ^{210}Pb activities within sediments exported from the basin via the basin mouth. The residence time and radionuclide inventory in each box is then determined using a series of equations that estimate the input and output fluxes from each box. The model developed by Dominik et al. (1987) assumed that the transfer of FRNs was entirely in association with particulate matter, but the model was later revised by Le Cloarec et al. (2007) to account for the transport of dissolved radionuclides through the system using established Kd values. Le Cloarec et al. (2007) suggest that uncertainty in the model results are primarily associated with errors in (1) the measurement of FRN fluxes out of the basin with suspended particulate matter, and (2) the determination of the atmospheric fluxes of ^{7}Be and ^{210}Pb. Perhaps a more important source of uncertainty is the assumption that ^{7}Be export from the soil box (upland areas) to the river box is negligible as it will be lost as a result of decay. This assumption contradicts the studies described earlier that suggest eroded surface sediments tagged with ^{7}Be can be used to determine the age of the sediment relative to the time it enters the channel.

Less uncertainty may be involved with the use of FRNs to assess the relative timing and magnitude of deposition and mobilization during flood events on the

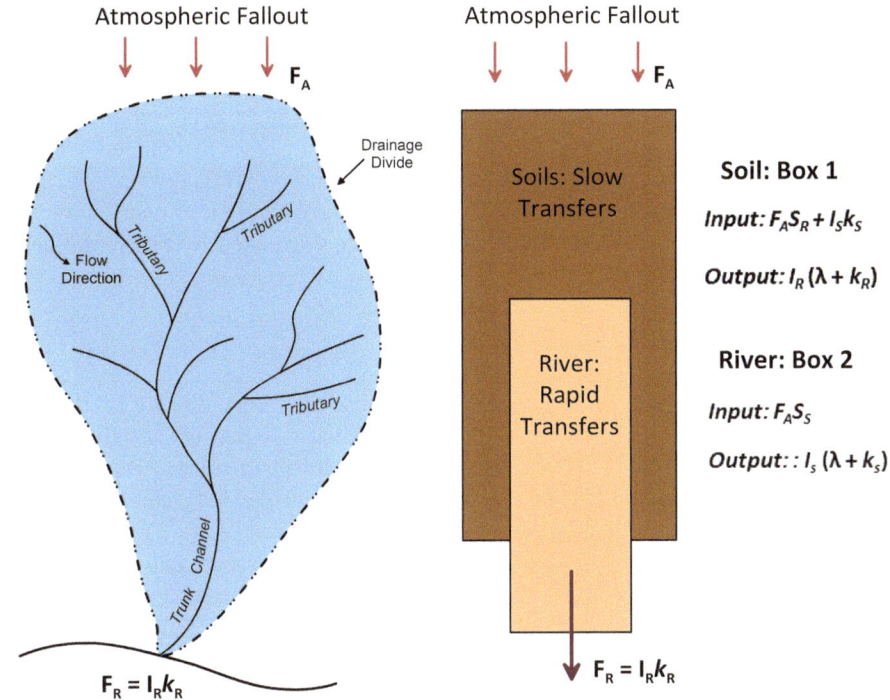

Fig. 3.10 Two-box radionuclide mass-balance model developed by Dominik et al. (1987) and modified by Le Cloarec et al. (2007) to assess sediment residence times in soil and river channels. FA atmospheric fallout flux; SS area of soil; SR area of river; IS radioactive inventory in soil box; IR radioactive inventory in river box; FR flux/export from river box (catchment); λ constant of radioactive decay; kS rate of output from soil box; kR rate of output from river (modified from Evrard et al. 2010)

basis of the vertical distribution of ^{7}Be within the channel bed material. Fitzgerald et al. (2001), for example, used ^{7}Be depth profiles to examine the amount of sediment deposition and resuspension along a section of the Fox River in Wisconsin where the resuspension of contaminated bed sediments is thought to be the primary source of PCBs to the overlying water column and Green Bay. Fitzgerald et al. (2001) argue that changes in ^{7}Be activity with depth are indicative of the timing and nature of scour and fill. For a majority of the cores obtained from the channel bed ^{7}Be decreased with depth. This downward decreasing trend in ^{7}Be activity is likely to result from the continued deposition of ^{7}Be tagged sediment that, upon isolation from additional atmospheric inputs, decays. Thus, the older sediments with the lowest activities are at the bottom of the profile. In contrast, a peak in ^{7}Be activity at depth may be produced by the slow deposition and mixing of ^{7}Be tagged sediment, followed by the rapid deposition of diluted ^{7}Be containing sediment that buries enriched ^{7}Be layer. The break in ^{7}Be activity can therefore be used to identify differences in the rate of deposition, and potentially, the amount of deposition that occurred during an individual event.

Fitzgerald et al. (2001) were also able to quantify the rates of deposition and resuspension by comparing the new inventory in the sediment core to the residual inventory. The residual inventory was taken to be the ^{7}Be activity within the core during an earlier sampling period (decay-corrected to the more recent sampling date), whereas the new inventory was calculated as the total inventory (obtained from the most recent sampling campaign) less the residual inventory. When the new inventory exceeded the residual inventory, deposition had occurred and vice versa. Moreover, short-term depositional or erosional rates, Ψ, were calculated by Fitzgerald et al. (2001) using the equation:

$$\Psi = \frac{fl_{7Be}}{\mu_{7Be}} \tag{3.7}$$

where fl_{7Be} is the ^{7}Be flux equal to the new deposit inventory divided by the time between the sampling intervals and μ_{7Be} is the mean ^{7}Be particle activity in suspended sediments. In the study by Fitzgerald et al. (2001), the mean ^{7}Be activity in suspended particles was determined by the analysis of suspended particles captured in sediment traps.

Fisher et al. (2010) also noted that important insights into scour and fill processes could be gained by examining the vertical ^{7}Be activity in channel bar deposits. In fact, they developed a conceptual model, presented in Fig. 3.11, that illustrates the types of vertical patterns that would result from differing sequences of scour and

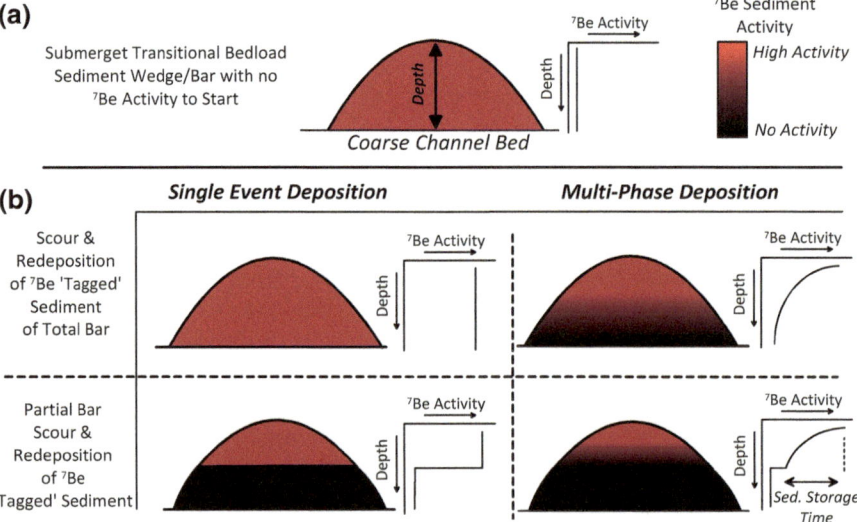

Fig. 3.11 Conceptual model of sediment bar scour and deposition as determined from ^{7}Be depth profiles. **a** Submerged bar composed of sediment with no ^{7}Be activity. *Solid black line* on activity versus depth plot shows that ^{7}Be activity is below detection throughout bar. **b** Typical depth-activity profiles observed for single and multi-phase depositional events. Change in ^{7}Be activity with depth is related to time of sediment storage in bar (modified from Fisher et al. 2010)

fill. Quantitatively, they calculate sedimentation rates by rearranging the CIA model described above such that (Appleby and Oldfield 1992):

$$A = A_0 e^{-\frac{\lambda m}{r}} \tag{3.8}$$

where r is the mass flux, given in g cm^{-2} d^{-1}, and m is the cumulative dry mass of sediment per unit area above the lower boundary (g cm^{-2}) (i.e., above the buried layer). This particular model assumes a constant initial concentration of ^{7}Be such that its rate of accumulation is proportional to the mass of sediment accumulated. The ^{7}Be activity in the cores, then, should decrease with depth monotonically where this assumption is met. Fisher et al. (2010) note that care should be taken when applying the model as its results represent short-term depositional rates that are constrained by the assumptions inherent in the CIA model, and are not necessarily indicative of erosional or depositional rates that occur at other sites spread across the channel bed.

Although it is widely accepted that vertical trends in ^{7}Be activity can provide important insights into the time and nature of sediment accumulation and remobilization processes, the resulting trends may be complicated by grain filtration along coarse (gravel) bed streams. Grain filtration is the process in which fine particles are moved downward into the interstitial pore spaces of a stable, coarser sedimentary layer at the channel surface (Gooseff et al. 2006; Haynes et al. 2009). In a study of the regulated and unregulated streams in Vermont and New Hampshire, Gartner et al. (2012) found that ^{7}Be activity in some cores varied in a non-systematic manner with depth. They interpreted these profiles to result from channel bed scour and fill processes similar to that suggested by Fitzgerald et al. (2001) in which higher activity layers could be buried by low activity layers (and vice versa); the two layers consisting of sediment from different sources. However, Gartner et al. (2012) suggested that downward decreasing ^{7}Be activities observed in a number of their cores was due to filtration where ^{7}Be activity was attenuated with increasing depth, rather than the continuous accumulation and burial of ^{7}Be tagged sediment. Thus, the systematic downward decrease in activity can result from both a continuous accumulation of sediment and/or filtration. Exactly which process is of most importance can probably be deciphered for most rivers on the basis of the stability and grain size of the bed material. Filtration is likely to be predominant along rivers with coarse-grained bed material that is relatively stable during low- to moderate-flood events.

3.5 Use of Fallout Radionuclides as an Age Dating Tool

Fluvial deposits associated with floodplains, terraces, riparian wetlands, and reservoirs contain an historical record, often extending over hundreds to thousands of years, of sediment source, production, transport, and storage within a catchment. While this archive is often discontinuous and incomplete, it can be effectively unraveled to gain insights into a wide range of environmental problems such as the effects of natural and anthropogenic disturbances (deforestation, wildfires, urban development, agriculture, etc.) on erosional and depositional processes, or the history of contaminant loadings to a river system by both point- and non-point sources of contamination. In order to construct and decipher changes in the alluvial record, the sediments must be accurately dated. As a result, a wide variety of methods have been developed to date sedimentary deposits found in riverine environments. Each of the developed methods differs in the time scale to which they are applicable, the accuracy and precision of the age estimates, and the nature of the depositional process that is being dated (see Stokes and Walling 2003 for a review). The use of fallout radionuclides (predominantly ^{137}Cs and ^{210}Pb) represents one of the most important methods to date historic riverine sediments (i.e., those $<\sim150$ years old). In the interest of space, we will not dive into the use of ^{137}Cs and ^{210}Pb as a dating tool as their applications have been summarized in a number of other documents (e.g., Appleby and Oldfield 1978; Robbins 1978; Appleby et al. 1979; Appleby 2001; He and Walling 1996; Stokes and Walling 2003; Belyaev et al. 2013; Golosov et al. 2012; Walling 2013).

References

Appleby P (2001) Chronostratigraphic techniques in recent sediments. In: Last W, Smol J (eds) Tracking environmental change using lake sediments, developments in paleoenvironmental research, vol 1. Springer, Netherlands, pp 171–203

Appleby P, Oldfield F (1978) The calculation of lead-210 dates assuming a constant rate of supply of unsupported ^{210}Pb to the sediment. Catena 5:1–8

Appleby P, Oldfield F (1992) Applications of ^{210}Pb to sedimentation studies. Oxford Science, Oxford, pp 731–778

Appleby PG, Oldfield F, Thompson R, Huttenen P, Tolonen K (1979) ^{210}Pb dating of annually laminated lake sediments from Finland. Nature 280:53–55

Basher L, Matthews K (1993) Relationship between ^{137}Cs in some undisturbed New Zealand soils and rainfall. Soil Res 31:655–663

Baskaran M (2011) Environmental isotope geochemistry: past, present, and future. In: Baskaran M (ed) Handbook of environmental isotope geochemistry. Advances in Isotope Geochemistry. Springer, Berlin, pp 3–10

Baskaran M, Coleman C, Santschi P (1993) Atmospheric depositional fluxes of ^{7}Be and ^{210}Pb at Galveston and college station, Texas. J Geophys Res-Atmos 98:20555–20571

Belyaev V, Golosov V, Markelov M, Evrard O, Ivanova N, Paramonova T, Shamshurina E (2013) Using Chernobyl-derived ^{137}Cs to document recent sediment deposition rates on the River Plava floodplain (Central European Russia). Hydrol Process 27:807–821

Bettoli M, Cantelli L, Degetto S, Tositti L, Tubertini O, Valcher S (1995) Preliminary investigations on ^{7}Be as a tracer in the study of environmental processes. J Radioanal Nucl Chem 190:137–147

Blake WH, Walling DE, He Q (1999) Fallout ^{7}Be as a tracer in soil erosion investigations. Appl Radiat Isot 51:599–605

Blake W, Wallbrink P, Wilkinson S, Humphreys G, Doerr S, Shakesby R, Tomkins K (2009) Deriving hillslope sediment budgets in wildfire-affected forests using fallout radionuclide tracers. Geomorphology 104:105–116

Bonniwell E, Matisoff G, Whiting P (1999) Fine sediment residence times in rivers determined using fallout radionuclides (^{7}Be, ^{137}Cs, ^{210}Pb). Geomorphology 27:75–92

Brigham M, McCullough C, Wilkinson P (2001) Analysis of suspended-sediment concentrations and radioisotope levels in the Wild Rice River basin, Northwestern Minnesota, 1973–98. US Department of the Interior, US Geological Survey

Bunzl K, Hotzl H, Rosner G, Winkler R (1995) Unexpectedly slow decrease of Chernobyl-derived radiocesium in air and deposition in Bavaria, Germany. Naturwissenschaften 82:417–420

Caitcheon G, Olley J, Pantus F, Hancock G, Leslie C (2012) The dominant erosion processes supplying fine sediment to three major rivers in tropical Australia, the Daly (NT), Mitchell (Qld) and Flinders (Qld) rivers. Geomorphology 151–152:188–195

Ceaglio E, Meusburger K, Freppaz M, Zanini E, Alewell C (2012) Estimation of soil redistribution rates due to snow cover related processes in a mountainous area, Valle d'Aosta, NW Italy. Hydrol Earth Syst Sci 16:517–528

Cushing C, Minshall G, Newbold J (1993) Transport dynamics of fine particulate organic matter in two Idaho streams. Limnol Oceanogr 38:1101–1101

Dickin A (1997) Radiogenic isotope geology. Cambridge University Press, Cambridge

Dietrich W, Dunne T (1978) Sediment budget for a small catchment in mountainous terrain. Z Geomorphol 29:191–206

Di Stefano C, Ferro V, Porto P (1999) Linking sediment yield and caesium-137 spatial distribution at basin scale. J Agric Eng Res 74:41–62

Dominik J, Burrus D, Vernet J (1987) Transport of the environmental radionuclides in an alpine watershed. Earth Planet Sci Lett 84:165–180

Dong W, Tims S, Fifield L, Guo Q (2010) Concentration and characterization of plutonium in soils of Hubei in central China. J Environ Radioact 101:29–32

Dörr H, Münnich K (1987) Spatial distribution of soil-^{137}Cs and ^{134}Cs in West Germany after Chernobyl. Naturwissenschaften 74:249–251

Elliott G, Campbell B, Loughran R (1990) Correlation of erosion measurements and soil caesium-137 content. Intel J Appl Radiat Is 41:713–717

Everett S, Tims S, Hancock G, Bartley R, Fifield L (2008) Comparison of Pu and ^{137}Cs as tracers of soil and sediment transport in a terrestrial environment. J Environ Radioact 99:383–393

Evrard O, Nemery J, Gratiot N, Duvert C, Ayrault S, Lefevre I et al. (2010) Sediment dynamics during the rainy season in tropical highland catchments of central Mexico using fallout radionuclides. Geomorphology 124:42–54

Fisher G, Magilligan F, Kaste J, Nislow K (2010) Constraining the timescales of sediment sequestration associated with large woody debris using cosmogenic ^{7}Be. J Geophys Res-Earth Surface 115:F01013

Fitzgerald SA, Klump JV, Swarzenski PW, Mackenzie RA, Richards KD (2001) Beryllium-7 as a tracer of short-term sediment deposition and resuspension in the Fox River, Wisconsin. Environ Sci Technol 35:300–305

Fornes W, Whiting P, Wilson C, Matisoff G (2005) Caesium-137-derived erosion rates in an agricultural setting: the effects of model assumptions and management practices. Earth Surf Proc Land 30:1181–1189

Foster IDL, Dalgeish H, Dearing JA, Jones ED (1994) Quantifying soil erosion and sediment transport in drainage basins: some observations on the use of ^{137}Cs, vol 224. IAHS Press, Wallingford

Fredericks D, Perrens S (1988) Estimating erosion using caesium-137: II. Estimating rates of soil loss. Sediment budgets, vol 174. IAHS Publication, Wallingford, pp 233–240

Gartner JD, Renshaw CE, Dade WB, Magilligan FJ (2012) Time and depth scales of fine sediment delivery into gravel stream beds: constraints from fallout radionuclides on fine sediment residence time and delivery. Geomorphology 151–152:39–49

Gaspar L, Navas A, Walling D, Machín J, Gómez Arozamena J (2013) Using ^{137}Cs and ^{210}Pb$_{ex}$ to assess soil redistribution on slopes at different temporal scales. Catena 102:46–54

Golosov V, Panin A, Markelov M (1999) Chernobyl ^{137}Cs redistribution in the small basin of the Lokna river, Central Russia. Phys Chem Earth Part A 24:881–885

Golosov V, Belyaev V, Markelov M (2013) Application of Chernobyl-derived ^{137}Cs fallout for sediment redistribution studies: lessons from European Russia. Hydrol Process 27:781–794

Gooseff M, Anderson J, Wondzell S, LaNier J, Haggerty R (2006) A modeling study of hyporheic exchange pattern and the sequence, size, and spacing of stream bedforms in mountain stream networks, Oregon, USA. Hydrol Process 20:2443–2457

Haynes H, Vignaga E, Holmes W (2009) Using magnetic resonance imaging for experimental analysis of fine-sediment infiltration into gravel beds. Sedimentology 56:1961–1975

He Q, Walling D (1996) Use of fallout Pb-210 measurements to investigate longer-term rates and patterns of overbank sediment deposition on the floodplains of lowland rivers. Earth Surf Proc Land 21:141–154

Hoefs J (2010) Geochemical fingerprints: a critical appraisal. Eur J Miner 22:3–15

Hoo WT, Fifield LK, Tims SG, Fujioka T, Mueller N (2011) Using fallout plutonium as a probe for erosion assessment. J Environ Radioact 102:937–942

Huisman N, Karthikeyan K (2012) Using radiometric tools to track sediment and phosphorus movement in an agricultural watershed. J Hydrol 450–451:219–229

Kachanoski R (1987) Comparison of measured soil 137-cesium losses and erosion rates. Can J Soil Sci 67:199–203

Kasprak A, Magilligan F, Nislow K, Renshaw C, Snyder N, Dade W (2013) Differentiating the relative importance of land cover change and geomorphic processes on fine sediment sequestration in a logged watershed. Geomorphology 185:67–77

Kaste J, Fernandez I, Hess C, Norton S (1999) Delivery of cosmogenic beryllium-7 to forested watersheds in Maine, USA. Geol Soc Am Abstr Programs 31:305

Kaste J, Norton S, Hess C (2002) Environmental chemistry of beryllium-7. Rev Miner Geochem 50:271–289

Kaste J, Elmore A, Vest K, Okin G (2011) Beryllium-7 in soils and vegetation along an arid precipitation gradient in Owens valley, California. Geophys Res Lett 38(L09):401

Kelsey H, Madej M, Pitlick J, Stroud M, Coghlan P (1981) Major sediment sources and limits to the effectiveness of erosion control techniques in the highly erosive watersheds of North Coastal California. In: Proceedings of a symposium on erosion and sediment transport in Pacific Rim Steeplands, vol 132. IAHS Publication, pp 93–510

Ketterer ME, Zhang J, Yamada M (2011) Application of transuranics as tracers and chronometers in the environment. In: Baskaran M (ed) Handbook of environmental isotope geochemistry. Advance in Isotope Geochemistry. Springer, Berlin, p 571

Koch D, Jacob D, Graustein W (1996) Vertical transport of tropospheric aerosols as indicated by [7]Be and ^{210}Pb in a chemical tracer model. J Geophys Res-Atmos 101:18651–18666

Kwam Kim J, Onda Y, Yang D, Kim M (2013) Temporal variations of reservoir sediment sources in a small mountainous catchment in Korea. Earth Surf Proc Land 38:1380–1392

Lance J, McIntyre S, Naney J, Rousseva S (1986) Measuring sediment movement at low erosion rates using cesium-137. Soil Sci Soc Am J 50:1303–1309

Le Cloarec M, Bonté P, Lefèvre I, Mouchel J, Colbert S (2007) Distribution of [7]Be, ^{210}Pb and ^{137}Cs in watersheds of different scales in the Seine River basin: inventories and residence times. Sci Total Environ 375:125–139

Livens FR, Howe MT, Hemingway JD, Goulding KWT, Howard BJ (1996) Forms and rates of release of ^{137}Cs in two peat soils. European J of Soil Sci 47:105–112

Lomenick T, Tamura T (1965) Naturally occurring fixation of cesium-137 on sediments of Lacustrine origin. Soil Sci Soc Am J 29:383–387

Longmore M (1982) The caesium-137 dating technique and associated applications in Australia—a review. In: Archaeometry: an australasion perspective, pp 310–321

Mabit L, Bernard C, Laverdiére M (2002) Quantification of soil redistribution and sediment budget in a Canadian watershed from fallout caesium-137 (^{137}Cs) data. Can J Soil Sci 82:423–431

Mabit L, Benmansour M, Walling D (2008) Comparative advantages and limitations of the fallout radionuclides ^{137}Cs, [210]Pb$_{ex}$ and [7]Be for assessing soil erosion and sedimentation. J Environ Radioact 99:1799–1807

Mabit L, Meusburger K, Fulajtar E, Alewell C (2013) The usefulness of ^{137}Cs as a tracer for soil erosion assessment: a critical reply to Parsons and Foster (2011). Earth Sci Rev 127:300–307

Matisoff G, Whiting P (2011) Measuring soil erosion rates using natural ([7]Be, ^{210}Pb) and anthropogenic (^{137}Cs, [239,240]Pu) radionuclides. In: Baskaran M (ed) Handbook of environmental isotope geochemistry. Advances in Isotope Geochemistry. Springer, Berlin, pp 487–519

Matisoff G, Wilson C, Whiting P (2005) The ^{7}Be/^{210}Pb − xs ratio as an indicator of suspended sediment age or fraction new sediment in suspension. Earth Surf Proc Land 30:1191–1201

Menzel R (1960) Transport of strontium-90 in runoff. Science 131:499–500

Nagle G, Fahey T, Ritchie J, Woodbury P (2007) Variations in sediment sources and yields in the Finger Lakes and Catskills regions of New York. Hydrol Process 21:828–838

Nouira A, Sayouty E, Benmansour M (2003) Use of ^{137}Cs technique for soil erosion study in the agricultural region of Casablanca in Morocco. J Environ Radioact 68:11–26

Olley J, Burton J, Smolders K, Pantus F, Pietsch T (2013) The application of fallout radionuclides to determine the dominant erosion process in water supply catchments of subtropical South-East Queensland, Australia. Hydrol Process 27:885–895

Parsons A, Foster I (2011) What can we learn about soil erosion from the use of ^{137}Cs? Earth Sci Rev 108:101–113

Perkins R, Thomas C (1980) Worldwide fallout. In: Hanson WC (ed) Transuranic elements in the environment USDOE/TIC-22800. US Department of Energy, Washington, pp 53–82

Porcillie D, Baskaran M (2011) An overview of isotope geochemistry in environmental studies. In: Baskaran M (ed) Handbook of environmental isotope geochemistry. Advances in Isotope Geochemistry. Springer, Berlin

Porto P, Walling DE (2012) Validating the use of ^{137}Cs and ^{210}Pb$_{ex}$ measurements to estimate rates of soil loss from cultivated land in Southern Italy. J. Environ Radioact 106:47–57

Porto P, Walling D, Ferro V (2001) Validating the use of caesium-137 measurements to estimate soil erosion rates in a small drainage basin in Calabria, Southern Italy. J Hydrol 248:93–108

Porto P, Walling D, Ferro V, di Stefano C (2003a) Validating erosion rate estimates provided by caesium-137 measurements for two small forested catchments in Calabria, Southern Italy. Land Degrad Dev 14:389–408

Porto P, Walling D, Tamburino V, Callegari G (2003b) Relating caesium-137 and soil loss from cultivated land. Catena 53:303–326

Preiss N, Mélières M, Pourchet M (1996) A compilation of data on lead 210 concentration in surface air and fluxes at the air-surface and water-sediment interfaces. J Geophys Res Atmos 101:28847–28862

Reid L, Dunne T, Cederholm C (1981) Application of sediment budget studies to the evaluation of logging road impact. J Hydrol 20:49–62

Ritchie J, McHenry J (1990) Application of radioactive fallout cesium-137 for measuring soil erosion and sediment accumulation rates and patterns: a review. J. Environ Qual 19:215–233

Ritchie J, Ritchie C (2008) Bibliography of publications of ^{137}Cs studies related to erosion and sediment deposition. http://www.ars.usda.gov/Main/docs.htm?docid=15237

Ritchie J, Spraberry J, McHenry J (1974) Estimating soil erosion from the redistribution of fallout ^{137}Cs. Soil Sci Soc Am J 38:137–139

Robbins J (1978) Geochemical and geophysical applications of radioactive lead. The Biogeochemistry of lead in the environment. Elsevier, Amsterdam, pp 285–337

Rogowski A, Tamura T (1965) Environmental mobility of cesium-137. Radiat Bot 10:35–45

Saç MM, Uğur A, Yener G, Özden B (2008) Estimates of soil erosion using cesium-137 tracer models. Environ Monit Assess 136:461–467

Salant N, Renshaw C, Magilligan F, Kaste J, Nislow K, Heimsath A (2007) The use of short-lived radionuclides to quantify transitional bed material transport in a regulated river. Earth Surf Proc Land 32:509–524

Schuller P, Iroumé A, Walling D, Mancilla H, Castillo A, Trumper R (2006) Use of beryllium-7 to document soil redistribution following forest harvest operations. J Environ Qual 35:1756–1763

Smith H, Blake W, Owens P (2012) Application of sediment tracers to discriminate sediment sources following wildfire. IAHS-AISH Publication, Wallingford, pp 81–89

Soto J, Navas A (2008) A simple model of Cs-137 profile to estimate soil redistribution in cultivated stony soils. Radiat Meas 43:1285–1293

Staunton S (1994) Adsorption of radiocaesium on various soils: interpretation and consequences of the effects of soil: solution ratio and solution composition on the distribution coefficient. Eur J Soil Sci 45:409–418

Stokes S, Walling D (2003) Radiogenic and isotopic methods for the direct dating of fluvial sediments. Wiley, New York, pp 231–267

Sutherland R (1994) Spatial variability of ^{137}Cs and the influence of sampling on estimates of sediment redistribution. Catena 21:57–71

Syversen N, Øygarden L, Salbu B (2001) Cesium-134 as a tracer to study particle transport processes within a small catchment with a buffer zone. J Environ Qual 30:1771–1783

Taylor A, Blake W, Couldrick L, Keith-Roach M (2012) Sorption behaviour of beryllium-7 and implications for its use as a sediment tracer. Geoderma 187-188:16–23

Trimble S (1983) A sediment budget for Coon Creek basin in the Driftless Area, Wisconsin, 1853–1977. Am J Sci 283:454–474

Wallbrink P, Croke J (2002) A combined rainfall simulator and tracer approach to assess the role of best management practices in minimising sediment redistribution and loss in forests after harvesting. For Ecol Manag 170:217–232

Wallbrink P, Murray A (1993) Use of fallout radionuclides as indicators of erosion processes. Hydrol Process 7:297–304

Wallbrink P, Murray A (1994) Fallout of ^{7}Be in south eastern Australia. J Environ Radioact 25:213–228

Wallbrink P, Murray A (1996) Distribution and variability of ^{7}Be in soils under different surface cover conditions and its potential for describing soil redistribution processes. Water Resour Res 32:467–476

Wallbrink PJ, Olley JM, Murray AS (1994) Measuring soil movement using ^{137}Cs: implications of reference site variability. Intl Assoc Hydrol Sci Publ 224:95–105

Wallbrink P, Murray A, Olley J, Olive L (1998) Determining sources and transit times of suspended sediment in the Murrumbidgee river, New South Wales, Australia, using fallout ^{137}Cs and ^{210}Pb. Water Resour Res 34:879–887

Wallbrink P, Roddy B, Olley J (2002) A tracer budget quantifying soil redistribution on hillslopes after forest harvesting. Catena 47:179–201

Walling D (2013) Beryllium-7: the cinderella of fallout radionuclide sediment tracers? Hydrol Process 27:830–844

Walling D, He Q (1999) Using fallout lead-210 measurements to estimate soil erosion in cultivated land. Soil Sci Soc Am J 63:1404–1412

Walling D, Quine T (1992) The use of caesium-137 measurements in soil erosion surveys. In: Bogen J, Walling DE, Day TJ (eds) Erosion and sediment transport monitoring programmes in river basins. In: Proceedings of the international symposium on international association of hydrological sciences, vol 210, pp 143–152

Walling DE, Woodward JC (1995) Tracing sources of suspended sediment in river basins: a case study of the River Clum, Devon, UK. Mar Freshw Res 46:327–336

Walling D, Owens P, Leeks G (1999) Fingerprinting suspended sediment sources in the catchment of the River Ouse, Yorkshire, UK. Hydrol Process 13:955–975

Walling D, He Q, Appleby P (2002) Conversion models for use in soil-erosion, soil-redistribution, and sedimentation investigations. Kluwer Academic Publishers, Dordrecht

Walling D, Schuller P, Zhang Y, Iroumé A (2009) Extending the timescale for using beryllium-7 measurements to document soil redistribution by erosion. Water Resour Res 45:W02418

Walling D, Zhang Y, He Q (2011) Models for deriving estimates of erosion and deposition rates from fallout radionuclide (caesium-137, excess lead-210, and beryllium-7) measurements and the development of user friendly software for model implementation. In: Impact of soil conservation measures on erosion control and soil quality IAEA-TECDOC-1665, pp 11–33

Wendling L, Harsh J, Ward T, Palmer C, Hamilton M, Boyle J, Flury M (2005) Cesium desorption from illite as affected by exudates from Rhizosphere bacteria. Environ Sci Technol 39:4505–4512

Wilkinson S, Prosser I, Rustomji P, Read A (2009) Modelling and testing spatially distributed sediment budgets to relate erosion processes to sediment yields. Environ Model Softw 24:489–501

Wilson C, Matisoff G, Whiting P (2003) Short-term erosion rates from a ^{7}Be inventory balance. Earth Surf Proc Land 28:967–977

Wise S (1980) Caesium-137 and lead-210: a review of the techniques and some applications in geomorphology. In: Cullingford RA, Davidson DA, Lewin J (eds) Timescales in geomorphology. Wiley, London, pp 109–127

Wooldridge D (1965) Tracing soil particle movement with Fe-59. Soil Sci Soc Am J 29:469–472

Zhang X, Zhang Y (1995) Use of caesium-137 to investigate sediment sources in the Hekouzhen-Longmen basin of the middle Yellow River, China. In: Foster IDL (ed) Sediment and water quality in river catchments. Wiley, London, pp 353–362

Zhang X, Higgitt D, Walling D (1990) A preliminary assessment of the potential for using caesium-137 to estimate rates of soil erosion in the Loess Plateau of China. Hydrol Sci J 35:243–252

Zhang X, Walling D, Quine T, Wen A (1997) Use of reservoir deposits and caesium-137 measurements to investigate the erosional response of a small drainage basin in the rolling loess plateau region of China. Land Degrad Dev 8:1–16

Zhang X, Qi Y, Walling D, He X, Wen A, Fu J (2006) A preliminary assessment of the potential for using $^{210}Pb_{ex}$ measurement to estimate soil redistribution rates on cultivated slopes in the Sichuan Hilly basin of China. Catena 68:1–9

Chapter 4
Radiogenic Isotopes

Abstract Radiogenic isotopes have been widely used to assess an extensive range of geological processes. In this chapter, we focus on the use of three radiogenic isotope systems (Sr, Nd, and Pb) to determine the source of sediment and sediment-associated contaminants in riverine environments. We begin by examining the past and continuing use of Sr and Nd isotopes to determine the provenance of sediment at large spatial scales before exploring their potential use to track anthropogenically contaminated sediment at much smaller (local to regional) spatial scales. We then turn our attention to the use of Pb isotopes as a tracer of Pb contaminated sediments in riverine environments. Given the many and increasing ways in which radiogenic isotopes can be applied to environmental, geomorphic, and hydrologic issues, the discussion is not meant to be exhaustive. Rather, it is intended to provide an overview of the sorts of methodological approaches that have been used to address the sediment/contaminant source problem in riverine environments, and the strengths and weaknesses inherent in the approach.

Keywords Sr · Nd · Pb isotopes · Sediment provenance · Pb contamination

4.1 Introduction

In the previous chapter we briefly described the use of fallout radionuclides as environmental tracers to gain insights into the source and transport processes of sediment and sediment-associated contaminants in river systems. Other isotopes have also been extensively used to trace sediment and any associated contaminants in the near surface environment, including fluvial systems (see Hoefs 2009 and Allègre 2008 for an overview of isotope geology and analysis). Of particular importance have been various radiogenic isotopes. Radiogenic isotopes are the daughter products generated from the decay of a radioactive parent. Here we focus on three isotope systems, Sr, Nd, and Pb. Their use as effective tracers hinges on several inherent characteristics, including the fact that they (1) can be precisely and accurately measured, (2) tend to exhibit isotopic abundances that vary widely within geological materials, and (3) are not significantly fractionated by physical, chemical or biological processes that alter the isotopic abundance observed within a given source material as it is dispersed through the river system. Thus, the isotopic ratios (fingerprints) of the

© The Author(s) 2015   89
J.R. Miller et al., *Application of Geochemical Tracers to Fluvial Sediment*,
SpringerBriefs in Earth Sciences, DOI 10.1007/978-3-319-13221-1_4

source sediments will remain largely intact during the erosion, transport, deposition, and diagensis of the sediment allowing them to serve as highly effective tracers. In the case of Sr and Nd, their isotopes have proven to be particularly effective at determining sediment provenance over large areas (ranging up to and including the global scale for the tracking of dust). In contrast, Pb isotopes have not been extensively used to track sediment per se, but are primarily used to track Earth materials contaminated by Pb and other related trace metals/metalloids from a particular source.

4.2 Sr and Nd Isotopic Systems

4.2.1 Tracing Sediments and Other Geological Materials

The abundances of Sr and Nd isotopes vary significantly in geological materials, primarily as a function of lithology and age of the rocks. These variations have been extensively used by the geological community to determine the source and dispersal patterns of sediment within oceanic, coastal, atmospheric and riverine systems (Table 4.1) (Nelson and DePaolo 1988; McLennan 1989; Awwiller 1994; Calanchi et al. 1996). Sr possesses four naturally occurring isotopes, of which ^{87}Sr is radiogenic, being produced from the radioactive decay of ^{87}Rb (Table 4.1). Since isotopic measurements are most easily and precisely made as isotopic ratios (Banner 2004), Sr isotopic data are generally presented by dividing radiogenic ^{87}Sr by ^{86}Sr (^{87}Sr/^{86}Sr), the latter a stable, non-radiogenic isotope. The relative mass difference between the isotopes (^{87}Sr and ^{86}Sr in this case) is relatively small ($\sim 1\%$). Thus, mass-dependent fractionation is limited, and that which does occur during analysis is typically corrected for. Thus, fractionation during sediment dispersal from a source as well as during sample analysis can generally be considered negligible (Banner 2004).

The ^{87}Sr/^{86}Sr ratios measured within oceanic basalts fall within a relatively narrow range of values (from 0.7020 to 0.7070) and within an even more narrowly defined range for mid-oceanic basalts (0.7022–0.7045) (Allègre 2008) (Fig. 4.1). In marked contrast, the ^{87}Sr/^{86}Sr ratios of granites and gneisses, which make up the bulk of

Table 4.1 Decay schemes and data for selected radiogenic isotopes (modified from Banner 2004)

Radioactive parent	Radiogenic daughter	Decay mechanism	Half-Life λ (billions of years)	Decay Constant
^{87}Rb	^{87}Sr	Beta	48.8	1.42×10^{-11}
^{147}Sm	^{143}Nd	Alpha	106	6.54×10^{-12}
^{238}U[a]	^{234}U; ^{230}Th; ^{206}Pb	Alpha & Beta	4.468	1.551×10^{-10}
^{235}U[a]	^{231}Pa; ^{207}Pb	Alpha & Beta	0.704	9.8485×10^{-10}
^{232}Th	^{208}Pb	Alpha & Beta	14.010	4.9475×10^{-11}
^{138}La	^{138}Ce	Beta branched	310	2.24×10^{-12}
^{176}Lu	^{176}Hf	Beta	35.7	1.94×10^{-11}

[a] Decays to produce a radioactive daughter which eventually decays to stable radiogenic isotope

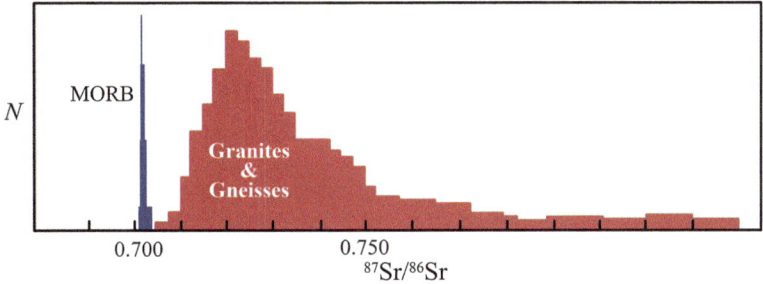

Fig. 4.1 Frequency diagram showing the distribution of ^{87}Sr/^{86}Sr ratios in mid-oceanic-ridge-basalts (MORBs) and granites and gneisses of the lower and upper continental crust (from Allègre 2008; Isotope Geology, Fig. 6.1, pp 222; copyright ©2008 Claude J Alle'gre. Reprinted with the permission of Cambridge University Press)

the continents, are highly variable, ranging from about 0.705 to more than 0.850 (Allègre 2008). The observed difference between the ^{87}Sr/^{86}Sr ratios of basalts and granites is dependent, in part, on the age of the rock units. Oceanic basalts exhibit an average age of about 80 Ma, the oldest being about 200 Ma (Allègre 2008). The age of the continental crust is highly variable as it consists of multiple tectonic segments sutured together in complex ways, but the age of the rock units can measure in the billions of years. Thus, the time over which ^{87}Rb in continental rocks has had to decay to ^{87}Sr, while variable, is typically much longer than for basalts, producing ^{87}Sr/^{86}Sr ratios in continental materials that are generally higher than those observed in basalt. The ^{87}Sr/^{86}Sr ratios also depend on the mineralogy and the initial Rb/Sr ratios within the rock units. As an alkaline earth element, the chemical characteristics of Sr are similar to that of Ca, and it often replaces Ca (as well as K) in rock forming minerals. Rb, on the other hand, is an alkali element and exhibits chemical properties similar to K. It is therefore more abundant as a replacement for K in K-rich minerals. Minerals that possess high K/Ca ratios are therefore likely to also possess higher Rb/Sr ratios, and through time will develop more radiogenic ^{87}Sr values (and vice versa). Put differently, for rocks containing minerals of similar age, those with high Rb and low Sr concentrations (e.g., biotite, muscovite) will be characterized by higher ^{87}Sr/^{86}Sr ratios than minerals possessing relative low Rb and high Sr contents (e.g., plagioclase, apatite) (Tripathy et al. 2011).

Nd, a light rare earth element, has seven naturally occurring isotopes. Two of these isotopes (^{142}Nd and ^{143}Nd) are radiogenic, produced by the radioactive decay of ^{146}Sm and ^{147}Sm, respectively. However, ^{146}Sm is an extinct radionuclide as it possesses a short half-life. Thus, tracer studies focus on ^{143}Nd/^{144}Nd. Unlike the isotopic abundances of Sr, isotopic Nd values are expressed in units of ε_{Nd} which portrays the deviation of ^{143}Nd/^{144}Nd in the sample relative to the ^{143}Nd/^{144}Nd in a standard. The use of ε makes it easier to report and interpret the data since the differences in ^{143}Nd/^{144}Nd values are very small. In this case, the utilized standard is the chondritic uniform reservoir. Mathematically,

$$\varepsilon_{Nd} = \left(\frac{(^{143}Nd/^{144}Nd)_{sample} - (^{143}Nd/^{144}Nd)_{chondrite}}{(^{143}Nd/^{144}Nd)_{chondrite}} \right) \times 10^4 \qquad (4.1)$$

As was the case of $^{87}Sr/^{86}Sr$, the $^{143}Nd/^{144}Nd$ ratio of rocks and minerals increases with time as a result of the production of ^{143}Nd from ^{147}Sm; thus, $^{143}Nd/^{144}Nd$ ratio (ε_{Nd}) depends on both the initial Sm/Nd ratio and age of the source rocks. In contrast to $^{87}Sr/^{86}Sr$, the highest (most radiogenic) $^{143}Nd/^{144}Nd$ ratios are found in mid-oceanic-basalts whereas the lowest ratios are associated with granites. Once again, however, the range of $^{143}Nd/^{144}Nd$ ratios are limited in basalts (about 0.5128–0.5120) but vary widely in continental crustal rocks from around 0.5080–0.5110 (Allègre 2008). Moreover, $^{87}Sr/^{86}Sr$ and $^{143}Nd/^{144}Nd$ ratios are inversely correlated indicating that there is a high degree of coherence between the Sr and Nd isotopes (Allègre 2008).

The wide variations in both $^{87}Sr/^{86}Sr$ and $^{143}Nd/^{144}Nd$ within geological materials, their highly conservative behavior, and the limited degree to which they are fractionated, suggests that the Sr and Nd isotopes will provide a diagnostic fingerprint of the constituents within both the dissolved and particulate load of rivers (as well as other water bodies). With regards to the dissolved load, Sr isotopes have been most extensively utilized, primarily to determine the source of the constituents within the water and the rates of chemical weathering (Moon et al. 2007; Rai and Singh 2007; Tripathy et al. 2010; Voss et al. 2014). On a global scale, the long-residence time of Sr in ocean waters compared to oceanic mixing times suggests that (1) the Sr isotopic composition of authigenic marine precipitates will exhibit the same ratios as the water at the time of their formation, and (2) the Sr values within the marine sediments will exhibit a high degree of homogeneity throughout the world's oceans (Banner 2004). Thus, recorded temporal variations in $^{87}Sr/^{86}Sr$ ratios through time, which now extend back to the Precambrian, describe global changes in ocean chemistry. In contrast, the residence time of $^{143}Nd/^{144}Nd$ in ocean waters is shorter than that of $^{87}Sr/^{86}Sr$ and shorter than the time required for complete oceanic mixing. Thus, observed variations in the $^{143}Nd/^{144}Nd$ values in marine sediments are thought to reflect regional differences in ocean chemistry. Significant effort has been devoted in recent years to determining the processes or factors that account for the noted variations in both the Sr and Nd isotopic ratios in marine sediments, including hydrothermal circulation at mid-oceanic ridges, alterations in weathering rates and intensities associated with climate change and/or tectonic activity, shifts in the configuration of the continents, and a number of other processes (see Banner 2004 for a review). It has been argued, for example, that the preferential release of Sr by the chemical weathering of newly exposed rock surfaces (resulting from tectonics or glaciation) can alter the $^{87}Sr/^{86}Sr$ signal of river systems, and the input of the Sr ratios to the oceans, in some areas for periods of about 20,000 years. Thus, it may be possible to link glacial episodes (or other mineral exposure processes) to documented alterations in the oceanic $^{87}Sr/^{86}Sr$ record.

On a much smaller scale Sr and Nd isotopes have been used to assess changes in the source of dissolved constituents within individual catchments. Voss et al. (2014), for example, utilized a time series of $^{87}Sr/^{86}Sr$ data from the Fraser River of Canada to determine changes in the source of dissolved constituents within a relatively pristine environment. They found that the Fraser River was characterized by seasonal variations in $^{87}Sr/^{86}Sr$ ratios with higher values occurring during the spring and summer and lower values during fall and winter. The isotopic data were then utilized in a mixing model to show that the higher $^{87}Sr/^{86}Sr$ values were associated with enhanced chemical weathering fluxes from old sedimentary rock units within the headwaters of the catchment, in spite of the fact that the area contributed relatively little water to the river.

Our primary interest here is in the use of Sr and Nd isotopes as a tracer of sediment and contaminated sediment provenance. The use of Sr isotopes as a means of determining sediment provenance was first demonstrated by Dasch (1969) who attributed large differences in the spatial variations of $^{87}Sr/^{86}Sr$ within the north Atlantic to variations in sediment source. Since then, both Sr and Nd isotopic systems have been widely used as sediment tracers. Their utilization is driven in large part by their highly conservative behavior in sediments that allows the isotopic signature of the source materials to be preserved within the sediment throughout its erosion, transport, deposition, and subsequent diagensis (DePaolo 1981; Goldstein et al. 1984; Jones et al. 1994; Winter et al. 1997; Tripathy et al. 2011). The Sr, and to lesser extent, Nd, isotopic composition of sediments has, however, been shown to vary with the particle size distribution of the sediments in some (e.g., Dasch 1969; Grousett and Biscaye 2005), but not all (e.g., Padoan et al. 2011) studies. Thus, the isotopic ratios within the sampled material may be altered from that of the source rocks by weathering and hydraulic sorting processes. In general, $^{87}Sr/^{86}Sr$ ratios are thought to be more significantly influenced than Nd because the latter is more uniformly distributed among the minerals in the parent rock. Thus, the preferential dissolution of specific minerals will impact $^{87}Sr/^{86}Sr$ ratios more than $^{143}Nd/^{144}Nd$ ratios. Similarly, mineralogical differences produced by particle sorting are likely to influence $^{87}Sr/^{86}Sr$ ratios more than $^{143}Nd/^{144}Nd$ ratios because of the wider range of the former among the minerals. The more conservative behavior of Nd is often considered to make it a more robust tracer (Walter et al. 2000; Tripathy et al. 2011). In addition, the observed differences in the geochemical behavior of Sr and Nd, when combined with their known inverse correlation in source rocks, have led to their combined use to assess sediment provenance and transport patterns (Tripathy et al. 2011). In other words, coherence in the temporal and spatial patterns within the sampled material is taken as an indication that the source signatures are preserved allowing for a determination of sediment provenance.

In light of the above, it is not uncommon for sediment sources and sampled mixtures to be shown on bivariate $^{87}Sr/^{86}Sr$ -$^{143}Nd/^{144}Nd$ (ε_{Nd}) plots (Fig. 4.2). Samples with $^{87}Sr/^{86}Sr$ -Nd values that systematically plot along a path between two end-member sources can be viewed as containing a mixture of sediments from the two sources. The isotopic Sr and Nd ratio of the mixture can be expressed in terms of the two sources as (Allègre 2008):

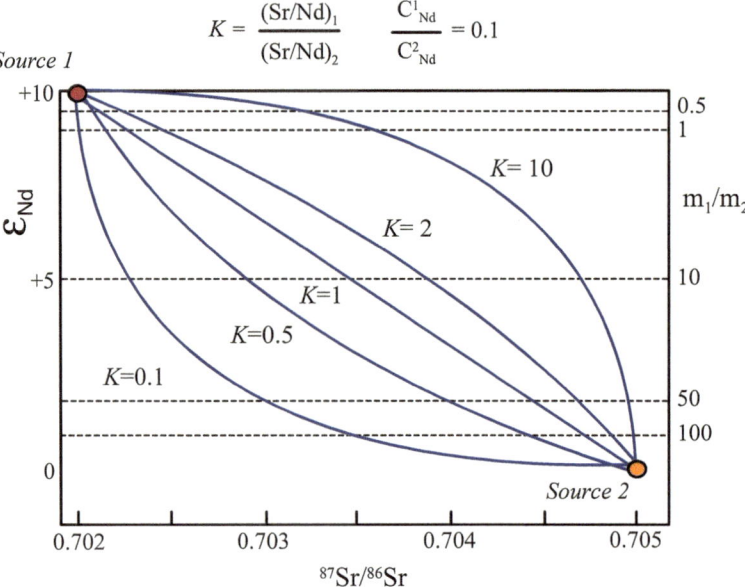

Fig. 4.2 Theoretical mixing *curves* plotted between two distinct end-member sources on a Nd-Sr diagram. *Curves* represent various relative concentrations of Nd and Sr within end members. K is the ratio of the Nd/Sr ratio in the two sources; the ratio of Nd concentrations in the sources is set at 0.1. The results are graduated in units of a mass ratio m_1/m_2 (from Allègre 2008; Isotope Geology, Fig. 6.17, pp 246; copyright ©2008 Claude J Alle'gre. Reprinted with the permission of Cambridge University Press)

$$R1_{mix} = R1_A x_1 + R1_B(1x_1) \tag{4.2}$$

$$R2_{mix} = R2_A y_1 + R2_B(1y_1) \tag{4.3}$$

where $R1$ and $R2$ are the isotopic ratios of interest (^{87}Sr/^{86}Sr ratio and ^{143}Nd/^{144}Nd) and

$$x_1 = \frac{m_1 C_A^1}{m_1 C_A^1 + m_2 C_B^2} \tag{4.4}$$

$$y_1 = \frac{m_1 C_B^1}{m_1 C_A^1 + m_2 C_B^2} \tag{4.5}$$

A and B are the two chemical elements of the ratios under consideration, C_A^1 and C_A^2 are the concentrations of element A in the two end-member sources, and C_B^1 and C_B^2 are the concentrations of element B in the two end-member sources. Equations 4.2 and 4.3 can be combined to produce a mixing equation that takes the form of:

$$\frac{R1_1 - R1_{mix}}{R1_{mix} - R1_2} = K \frac{R2_1 - R2_{mix}}{R2_{mix} - R2_2} \tag{4.6}$$

The term K in Eq. 4.6 has the value

$$K = \frac{C_A^1/C_B^1}{C_A^2/C_B^1} \tag{4.7}$$

In the case of Pb, as we will see below, the denominator of the two plotted ratios is the same; thus, the equation forms a straight mixing line between the two sources. A straight-line also is produced when the denominators differ (as is true for the use of $^{87}Sr/^{86}Sr$ -$^{143}Nd/^{144}Nd$), but K is equal to 1 indicating the Nd/Sr ratios are constant in both systems. However, when K is equal to any other negative or positive value, the line between the end-member sources forms a hyperbola, the shape of which varies with the value of K (Allègre 2008) (Fig. 4.2). Regardless of form, a systematic trend between two end-members allows for an estimation of the relative proportion of sediment from the two sources.

To date, Sr and Nd isotopes have primarily been used to determine the provenance of siliciclastic sedimentary rocks, particularly within sedimentary basins (McCulloch and Wasserburg 1978; Goldstein and Jacobsen 1988; Awwiller and Mack 1989; McLennan 1989; McLennan et al. 1990, 1993; Awwiller 1994). With regard to contemporary sediments, Sr and Nd isotopes have mainly been applied to large systems characterized by a diversity of rock types. Both Sr and Nd have, for example, been extensively utilized to assess the origins and transport patterns of dust (aerosols, loess and sand) on a worldwide basis (Grousset et al. 1992, 2005; Biscaye et al. 1997; Basile et al. 1997, 2001).

With regard to contemporary riverine systems, Sr-Nd fingerprints have often been applied to marine and coastal sediments to determine the relative contribution and dispersal patterns of sediment from river basins underlain by differing rock types and which therefore exhibit specific Sr-Nd isotopic signatures (Weldeab et al. 2002; Wei et al. 2012; Rosenbauer et al. 2013). Weldeab et al. (2002), for example, utilized Sr-Nd isotopic data to determine the source and primary transport pathways of suspended particulate matter within the Eastern Mediterranean Sea (Fig. 4.3). Similarly, Rosenbauer et al. (2013) applied Sr and Nd isotopes, along with selected trace elements and rare earth elements, to San Francisco Bay to determine the relative contributions of beach-sized sands from the major inflowing rivers. They found that the majority of the sand within three areas of the Bay (Suisun Bay, San Pablo Bay, and north Central Bay) originated from the Sierra Nevada Batholith via the Sacramento River, while input from other rivers including the Napa and San Joaquin provided lesser contributions. The Sr-Nd isotopic data also revealed that other sediment sources were locally important. Once the sand-sized particles exited the Bay, the isotopic data indicated that the materials were transported southward along the outer coast by long-shore currents (Rosenbauer et al. 2013).

The Sr-Nd signatures of sediment within individual tributaries have also been used to assess the primary sources and dispersal mechanics within large river basins. For instance, Padoan et al. (2011) used Sr-Nd isotopic ratios to determine the relative contributions of sediment input from the major tributaries to the Nile and the effects

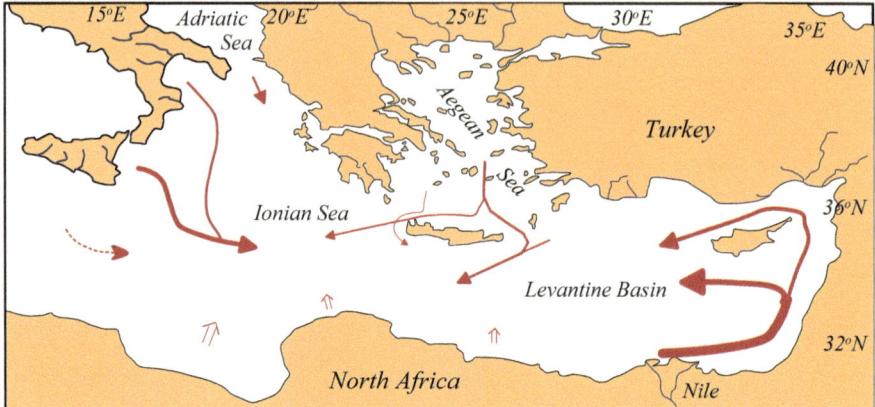

Fig. 4.3 The primary transport pathways of suspended riverine sediment within the Eastern Mediterranean Sea as determined from Sr and Nd isotopes. *Open arrows* indicate turbid flow directions as provided by Reeder et al. (1998) (from Weldeab et al. 2002)

of hydraulic sorting processes on the sediment size distribution. The Nile River Basin is well-suited for the study because individual catchments are underlain by rocks of different age and therefore exhibit different Sr-Nd isotopic ratios. More specifically, they found that tertiary volcanic rocks of the Ethiopian highlands, drained by the Blue Nile and Atbara (Fig. 4.4a), were characterized by low $^{87}Sr/^{86}Sr$ values and high $^{143}Nd/^{144}Nd$ ratios. Sediments from these rocks could be discriminated from Archean to Proterozoic crystalline basement rocks and Mesozoic sedimentary rocks exposed along the Western margin of the Red Sea and East African Rift (which were characterized by relatively high $^{87}Sr/^{86}Sr$ values and low $^{143}Nd/^{144}Nd$ ratios). When combined with isotopic data on samples from the main channel of the Nile, they were able to determine that the majority of the sediment was derived from the Blue Nile and Atbara. In spite of the fact that the White Nile supplies about 33 % of the discharge to the main Nile, and encompasses about 50 % of the total basin area, it contributes relatively little sediment to the main River (Fig. 4.4b). Apparently, very little sediment is transported through various East African Lakes and the vast wetlands of the Sudd and Machar marshes of South Sudan. The Sr-Nd isotopic data demonstrated that sediment fluxes to the lower Nile do not correspond to discharge or basin area, but are related to high rates of erosion in the Ethiopian highlands (Padoan et al. 2011).

4.2.2 Tracing Contaminated Particles

While the use of Sr and Nd isotopes to determine the source and dispersal patterns of sediment within siliciclastic strata, atmospheric systems, and large river-marine systems is widely recognized, their application to diffuse or point-sources

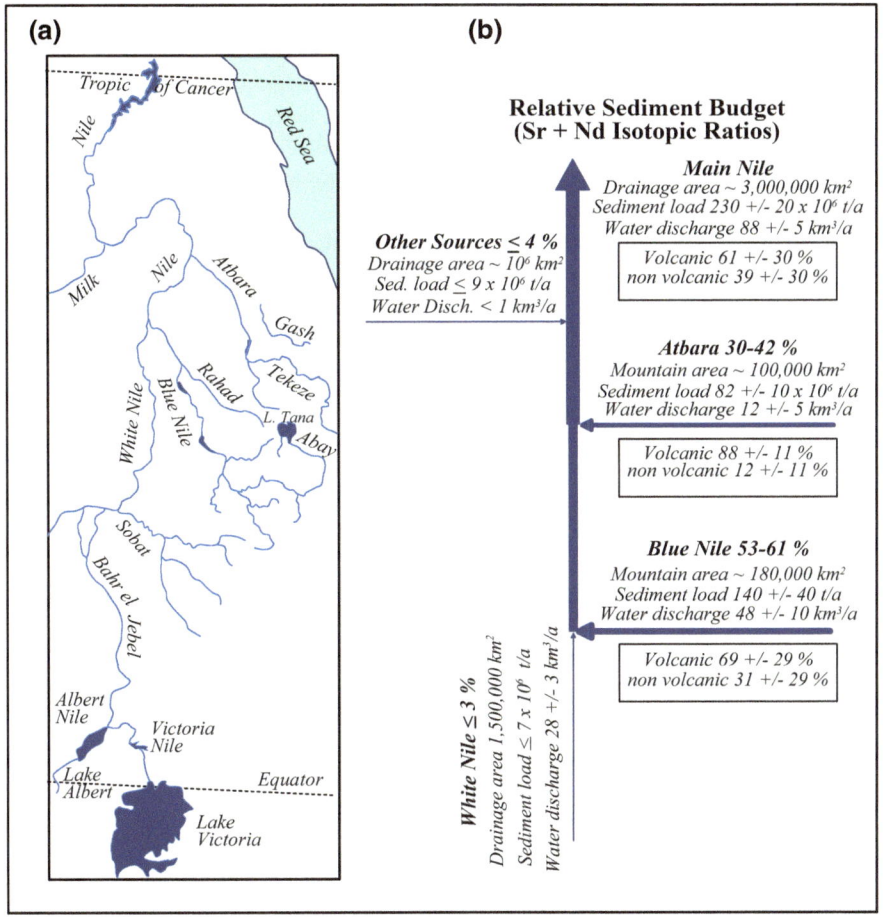

Fig. 4.4 a Map of the primary tributaries to the Nile River; **b** Relative sediment budget based on Sr and Nd isotopic provenance data. Relative sediment contributions were calculated using a inverse (mixing) model and the available isotopic data (modified from Padoan et al. 2011)

of contamination has thus far been limited, particularly with regards to rivers. At issue is whether Sr and Nd isotopic ratios vary to the extent needed to form an effective fingerprint. Nonetheless, a number of recent studies unrelated to rivers suggest that both isotopes can be used to discriminate between contaminated source materials and may therefore prove to be effective tracers of contaminated sediments in river systems (Lahd Geagea et al. 2007, 2008a, b; Kamenov et al. 2009; Guéguen et al. 2012). Lahd Geagea et al. (2008a), for example, demonstrated that particulate matter associated with urban pollutants within the area of Starsbourg-Kehl, Germany could be fingerprinted on the basis of Sr, Nd, and Pb isotopes (Table 4.2). Particularly large differences were found between the ε_{Nd} values of various industrial sources (e.g., waste incinerators and steel plants) and car soot, suggesting that they may serve as particularly powerful tracers of particulates from vehicle emissions, especially particulates from diesel engines (Lahd Geagea et al. 2008a). In addition, subsequent work by Guéguen et al. (2012) in the same area found that while chemical waste incinerators, domestic waste incinerators, thermal power plants, and steel plants exhibited similar Pb isotopic values and, therefore, could not be fingerprinted using only Pb isotopes, they could be defined using a combination of Pb, Sr, and Nd isotopes (Table 4.2). In fact, Sr and Nd isotopes were much more effective than Pb isotopes at differentiating the source of particulate matter from these industrial sources.

The potential to use Sr and Nd isotopes to document changes in chemical fluxes to aquatic systems has also been demonstrated. Kamenov et al. (2009), for example, examined the vertical (depth) variations in Sr, Nd, and Pb isotopes as well as selected trace metals and metalloids in a well-dated peat core from the Blue Cypress Marsh of southeast Florida. Geochemical changes in the composition of the sediments with depth were used to conclude that the flux of a number of toxic trace elements to the marsh from atmospheric sources had increased following European settlement. In addition, they found that significant systematic changes in both Sr concentrations and $^{87}Sr/^{86}Sr$ ratios occurred within the core (Fig. 4.5). Changes in $^{87}Sr/^{86}Sr$ values were consistent with, and attributed to, the influx of limestone dust from quarrying operations that were associated with urban development. Interestingly, the timing of the Sr isotopic shift in the core did not precisely correspond to the onset of quarrying, suggesting that Sr may have exhibited some slow downward diffusion within the peat deposits. Nd isotopes only exhibited a minor shift in ε_{Nd} values with depth in the core (at a depth of approximately 32–36 cm; Fig. 4.5). In contrast to Sr, the stratigraphic change in Nd isotopic values was attributed to variations in the relative

Table 4.2 Isotopic signatures of particulate matter from selected pollutant sources in the urban environment of Strasbourg-Kehl, Germany as reported by Lahd Geagea et al. (2008a) and Guéguen et al. (2012). Table modified from Guéguen et al. 2012

Pollutant source	$^{87}Sr/^{86}Sr$	$^{143}Nd/^{144}Nd$	$^{206}Pb/^{207}Pb$
Domestic waste incinerators[a]	0.70953	−9.7	1.1523[c]
Chemical waste incinerator[b]	0.71099	−8.4	1.1468[c]
Thermal power plants[a]	0.71241	−11.9	1.1528[c]
Steel plant[a]	0.70904	−17.5	1.1512[c]
Chimney soot[b]	0.71428	−11.2	1.1659
Soot—gasoline (2005)[a]	0.70881	−6.9	1.0898[c]
Soot—diesel (2005)[a]	0.70871	−6.0	1.1596[c]
Paper producer[b]	0.70992	−9.1	1.1621
Agricultural dust[b]	0.72326	−9.5	1.1798
Uncertainties 2σ	0.00001	0.4	0.00001
			0.0006[c]

[a] Lahd Geagea et al. (2008a)
[b] Guéguen et al. (2012)
[c] Measurements by TIMS; Other Pb isotopic measurements by MC-ICP-MS

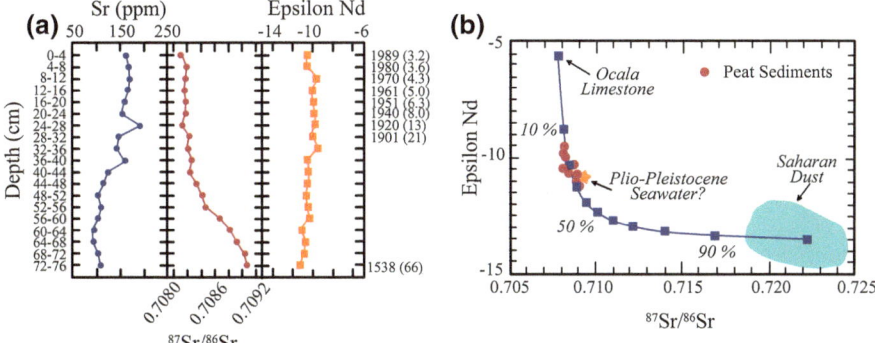

Fig. 4.5 a Variations in Sr concentration, ^{87}Sr/^{86}Sr ratios, and ε_{Nd} values as a function of depth below the ground surface and age within a peat deposit of southeast Florida. Age estimates based on ^{210}Pb and calibrated radiocarbon data from the 72–76 cm depth interval. **b** Bivariate ^{87}Sr/^{86}Sr versus ε_{Nd} plot showing mixing curve between suggested end-member sources (Saharan dust and Ocala limestone) and the peat deposits from southeast Florida. *Filled squares* represent relative mixing proportions between sources in terms of percent Saharan dust. See Kamenov et al. (2009) for parameters used to develop mixing curve. Plio-Pleistocene seawater value is based on Nd data for northeast Atlantic from Abouchami et al. (1999) (adapted from Kamenov et al. 2009)

contribution of dust from the Saharan. The argument is supported by the plotting of the sampled marsh sediments on a ε_{Nd} -^{87}Sr/^{86}Sr mixing curve (Fig. 4.5b) such that Nd isotopic variations can be explained by the mixing of 70–85 % sediment from the local bedrock and 15–30 % of the sediment from the Saharan.

The study by Kamenov et al. (2009) suggests that the use of Nd and Sr isotopes to track contaminant influx to aquatic systems may be more effective than their limited use for this purpose implies. Thus, the applications of Sr and Nd isotopes to contaminant problems in rivers and other aquatic systems is likely to increase in the future. They will probably be most effective when combined with other isotopes, such as those of Pb, or when used in combination with other forms of tracers (e.g., selected trace metals, metalloids, or Rare Earth Elements).

4.3 Pb Isotopes

4.3.1 General Characteristics

Pb has three radiogenic isotopes, ^{206}Pb, ^{207}Pb, and ^{208}Pb, which are derived from the radioactive decay of ^{238}U, ^{235}U, and ^{232}Th, respectively (Table 4.1). They are often combined in tracer studies with ^{204}Pb, which has no known radiogenic parent. Pb isotopes have not been extensively utilized to determine the source of sedimentary particles per se. However, it is fair to say that they now represent the method of choice for tracking Pb and Pb contaminated sediment. To date, they have been

Table 4.3 Pb isotopic abundance

Isotope	Abundance (%)
^{204}Pb	1.48
^{206}Pb	23.60
^{207}Pb	22.60
^{208}Pb	52.30

used to determine the source of Pb in air, aerosols, and dust (Ault et al. 1970; Chow and Johnstone 1965 ; Gulson et al. 1994, 1996; Rosman et al. 1993b; Chiaradia and Cupelin 2000; Laidlaw et al. 2014), snow and ice (Rosman et al. 1993a, b, 1997; Veysseyre et al. 2001; Shotyk et al. 2005), soils (Gulson et al. 1981; Steinmann and Stille 1997; Hansmann and Köppel 2000; Mihaljeviç et al. 2006; Reimann et al. 2012; Mackay et al. 2013; Kristensen et al. 2014), lacustrine and reservoir deposits (Shirahata et al. 1980; Petit et al. 1984; Chiaradia et al. 1997; Foster et al. 2002), wetlands and peat (Shotyk et al. 1998; Weiss et al. 1999; Marcantonio et al. 2000; Cloy et al. 2008), plants, mosses, and tree rings (Bellis et al. 2004; Bindler et al. 2004; Bi et al. 2009), human tissues and blood (Manton 1977; Keinonen 1992; Gulson et al. 1996) and other biota (Sangster et al. 2000; Miller et al. 2005; Soto-Jiménez et al. 2008, 2009; Sondergaard et al. 2010; Potot et al. 2012). Lead isotopes have also been extensively utilized to determine the origins of Pb and other correlated trace metals/metalloids in riverine environments, mostly from point-sources of contamination (see Bird 2011) for a detailed review of Pb as a contaminant tracer in rivers).

Like Sr, the isotopic abundances of Pb within geological materials are reported as ratios (e.g., ^{206}Pb/^{204}Pb). Unfortunately, investigators have not used a consistent set of ratios. While one study may report the values in terms of ^{206}Pb/^{207}Pb, others, particularly geologically oriented investigations, may report the same measured abundances in terms of ^{207}Pb/^{206}Pb. These differences in reporting make it more difficult than necessary to compare the results between investigations. Investigators have also focused on different ratios. For example, geologically oriented studies have extensively used the ratio of the radiogenic isotopes to ^{204}Pb (i.e., ^{206}Pb /^{204}Pb, ^{207}Pb/^{204}Pb, and ^{208}Pb/^{204}Pb) (Bird 2011). In contrast, most environmental studies only use the radiogenic isotopes of Pb as geochemical tracers for two reasons: (1) the abundance of ^{204}Pb is relatively low (Table 4.3) making it more difficult to accurately measure, particularly with an MC-ICP-MS, and (2) the discriminative (fingerprinting) power of Pb isotopes is primarily due to ^{206}Pb, ^{207}Pb, and ^{208}Pb (Sangster et al. 2000).

The effectiveness of Pb isotopes as a geochemical tracer is related to several factors. First, the radiogenic isotopes of Pb can be measured with a high degree of precision and accuracy. Second, effective contaminant tracers generally exhibit conservative behavior over a wide-range of environmental (geochemical) conditions. In the case of sediment, conservative behavior means that the tracer will move with the sediment without any significant loss in elemental mass (Yeager et al. 2005; Bird 2011). Moreover, fractionation of the radiogenic Pb isotopes by physical, chemical, and biological processes is limited because of the low relative atomic weight differences among the isotopes (Keinonen 1992; Komárek et al. 2008; Balcaen et al. 2010; Cheng and Hu 2010; Bird 2011). This is an extremely beneficial trait in that the Pb isotopic composition of the ore deposits does not change during mining, smelting, or other industrial or biological processing. Thus, the Pb released from a contaminant source retains the isotopic signature of the source material as it moves through the river system. As a result, the observed differences in Pb isotopic ratios between the source materials and the alluvial sediments at any given site is a function of the mixing with Pb in the sediments from the various sources that exist, allowing for a quantitative analysis of Pb provenance (Cheng and Hu 2010). Third, the most effective tracers exhibit a wide range of values in geological materials. In the case of Pb, the isotopic composition of a geological material depends upon the relative proportion of U, Th, and Pb in the system, various mixing processes associated with metamorphism, and the age of the rocks and minerals during which U and Th can decay to the radiogenic isotopes of Pb (Keinonen 1992; Cheng and Hu 2010). Although the Pb isotopic composition of ore deposits on a global scale is highly variable (Bird 2011), there is a strong tendency for ore-derived Pb to be more radiogenic. As a result, Pb ratios within anthropogenically derived materials created from those ores tend to differ from the more geogenic Pb typically found within non-mineralized rocks that underlie the basin (Hopper et al. 1991). It is often possible, then, to effectively fingerprint both natural and anthropogenic sources of Pb contained within alluvial sediments.

4.3.2 Applying Pb Isotopes as a Tracer in Riverine Environments

Although Pb isotopes have recently been used in riverine environments for a wide range of purposes (e.g., the elucidation of Pb cycling dynamics involved with sediment/rock/water interactions (Ip et al. 2007), they primarily have been used to determine contaminant source(s) or to construct a chronology of source/pollutant loading rates to the river. The analyses presume that the analyzed sediment is a mixture of particles from all of the potential Pb sources in the catchment, and that the relative abundance of the radiogenic Pb isotopes in the sediment reflects the contributions from each source. In addition, it is assumed that the isotopes behave conservatively with respect to physical processes (e.g., hydraulic sorting) that may lead to the partitioning of particles on the basis of size and/or density into specific sedimentologic units. While there is some evidence to suggest that physical sorting processes may lead to minor differences in the isotopic ratios from one location in the river to another (Bird 2011), it currently appears that differences related to such grain size affects are likely to be negligible along contaminated river systems. It follows, then, that the observed temporal and spatial variations in Pb isotopic ratios within alluvial sediments will predominately reflect the input and mixing of Pb from different natural and anthropogenic sources each of which are characterized by a different Pb isotopic composition.

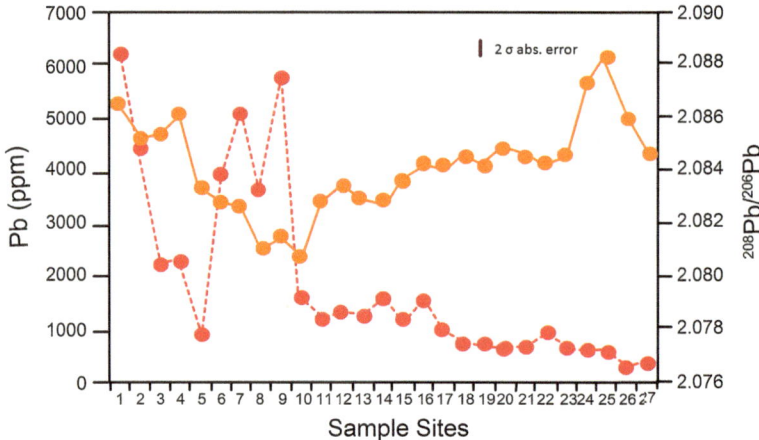

Fig. 4.6 Downstream changes in total Pb concentration (*red circles*) and $^{208}Pb/^{206}Pb$ ratios (*orange circles*) along the River Wear, Northern England. Sample sites 1 through 27 are shown on ordinal axial and cover a distance of 86.5 km. Error bar refers to isotopic measurements (from Shepherd et al. 2009)

An excellent example of the use of Pb isotopes to determine Pb provenance was provided by Shepherd et al. (2009) for the River Wear in northeast England. They analyzed a variety of geological materials within the catchment including the channel bed sediments along an 87 km reach of the river for selected Pb isotopes (Fig. 4.6). As is common, Pb concentration data were also collected to provide additional details on the levels of contamination associated with the various Pb sources (Fig. 4.6). At Site 1 the $^{208}Pb/^{206}Pb$ ratio is comparable to geological materials found in the outer mineralized zone of an orefield. The high Pb concentrations found in the channel bed sediments are therefore likely to reflect the influx of Pb-contaminated sediment to the river from historic mining operations, including abandoned mines, waste piles and an old processing plant that produced about 60, 000 t of Pb concentrate (Dunham 1990). The rapid downstream declines in Pb concentration from Sites 1–5 are thought to reflect dilution processes and the deposition of dense, finely milled Pb ore within the channel bed. Downstream of Site 5, Pb concentrations semi-systematically increase until reaching Site 9, whereas $^{208}Pb/^{206}Pb$ ratios continue to decline. The divergent changes in Pb concentration and $^{208}Pb/^{206}Pb$ ratios apparently reflect the influx of Pb-enriched mine waste from two tributary valleys where the Pb ore was derived from the inner zone of the orefield. Rocks within this inner zone exhibited a lower $^{208}Pb/^{206}Pb$ ratio than those found within the outer zone of the orefield, thereby lowering the ratio within the channel bed sediment. Between Sites 10 and 11 the $^{208}Pb/^{206}Pb$ ratio increases abruptly, reflecting the input of Pb from a tributary basin (Bollihope Beck) that contains another cluster of mines. Pb concentrations do not significantly increase indicating that the tributary provides only a limited quantity of the Pb to the axial channel of the River Wear (about 18 % of the total). There are no further direct inputs of Pb from mining operations downstream of Site 11. Without the isotopic data it could reasonably be argued that the decreasing Pb concentrations observed downstream of Site 11 was the result of dilution of sediment-born Pb by relatively 'clean' sediments from the underlying bedrock. However, the increasing $^{208}Pb/^{206}Pb$ ratio indicates that Pb characterized by a higher $^{208}Pb/^{206}Pb$ ratio is entering the channel from other source(s). Shepherd et al. (2009) argued that this additional Pb was associated with diffuse anthropogenic sources from tributaries draining an abandoned coalfield. The abrupt and sustained increase in $^{208}Pb/^{206}Pb$ between sites 24 and 27 was interpreted to result from either: (1) an increased influx of diffuse Pb from Durham; or (2) the influx of tetra-ethyl lead associated with leaded gasoline. The point to be made here is that the analysis by Shepherd et al. (2009) provides a much more detailed understanding of Pb sources and their relative contributions to the contamination of the river than would be possible by only examining the Pb concentration data.

The quantification of Pb contributions from a particular source is often conducted by combining observed spatial variations in isotopic ratios along a channel with the analysis of three-component (bivariate) scatter diagrams depicting differences between two isotopic ratios (Elbaz-Poulichet et al. 1986; Miller et al. 2002; McGill et al. 2003; Kurkjian et al. 2004; Bird et al. 2010a, b). When isotopic ratios of the analyzed samples form a linear trend, Pb within the samples is typically interpreted to be derived from two primary end-member sources (as was the case of Sr and Nd plots). In the case of Pb contaminated alluvial sediments, one end-member typically represents Pb from the underlying bedrock or the soils developed from it, whereas the other represents Pb from a significant anthropogenic source (Erel et al. 1997; Bird 2011). The signature of the 'geogenic' Pb source may be determined by analyzing a number of different materials including channel bed sediments from uncontaminated tributaries or areas upstream of the contaminant influx (Keinonen 1992), uncontaminated alluvial terraces (usually comprised of pre-historic sediments) (Miller et al. 2002, 2007; Church et al. 2004), alluvial sediment found at depth within a sediment core that pre-dates anthropogenic contamination (Church et al. 2004), or the direct analysis of the underlying bedrock (Miller et al. 2007) (Fig. 4.7).

Miller et al. (2007) provided an example of the use of bivariate scatter diagrams to assess the source and dispersal patterns of Pb along the Rio Pilcomayo downstream of the Cerro Rico de Potosi precious metal-polymetallic tin deposits of Bolivia. Mining and milling of the deposits has continuously occurred since 1545 and resulted in the severe contamination of the river by a wide-range of trace metals and metalloids (Hudson-Edwards et al. 2001; Miller et al. 2007). Miller et al. (2007) found that (1) bedrock units within the catchment exhibited different $^{206}Pb/^{207}Pb$ and $^{206}Pb/^{208}Pb$ ratios, and (2) alluvial sediments contained within pre-mining terrace deposits formed a linear mixing line in which the Ordovician and Mesozoic rocks served as the isotopic end-members (Fig. 4.8a). In marked contrast, samples collected in 2000 from the highly contaminated channel bed fall along a separate mixing line in which the isotopic end members were formed by Ordovician rocks and mine/mill processing waste from the operations at Cerro Rico (Fig. 4.8b). Samples collected in 2000, then, showed that the channel bed sediment was dominated by natural Pb from the underlying Ordovician rocks and Pb from the Cerro Rico ore deposits that were released from processing plants into the river system. Pb input from the Mesozoic rocks was apparently minor as the isotopic signature was masked by large inputs from the other two sources.

One of the significant advantages of using Pb isotopes as tracers is that accurate estimates of the relative contributions of Pb can be determined when the data plot along a mixing line (and other sources of Pb are assumed to be negligible) (Kristensen et al. 2014). Since the trend is linear, the estimated contributions are typically calculated using a simple binary model (Monna et al. 1997; Bird et al. 2010a, b; 2011) that takes the form of:

$$\%A = \frac{(^{20x}\mathrm{Pb}/^{20y}\mathrm{Pb})_S - (^{20x}\mathrm{Pb}/^{20y}\mathrm{Pb})_A}{(^{20x}\mathrm{Pb}/^{20y}\mathrm{Pb})_B - (^{20x}\mathrm{Pb}/^{20y}\mathrm{Pb})_A} \qquad (4.8)$$

where $\%A$ is the contribution of source A in a sample and $^{20x}\mathrm{Pb}/^{20y}\mathrm{Pb}_S$, $^{20x}\mathrm{Pb}/^{20y}\mathrm{Pb}_A$ and $^{20x}\mathrm{Pb}/^{20y}\mathrm{Pb}_B$ are the average ratios in the sample (S) and the two end-member sources, A and B, respectively. In the case of the Rio Pilcomayo, Miller et al. (2007) were able to use the equation to estimate the relative contribution of Pb from the upstream mining operations to the channel bed at selected locations along the river. The estimates for samples collected in 2000 ranged from >99 to about 56 % of the total Pb present (these estimates were recalculated from Miller et al. (2002) on the basis of a more recent and accurate understanding of the isotopic signatures of the ore deposits).

In addition to the spatial (downstream) variations in the contribution of Pb from the mines to the Rio Pilcomayo, Miller et al. (2007) found that the Pb isotopic ratios

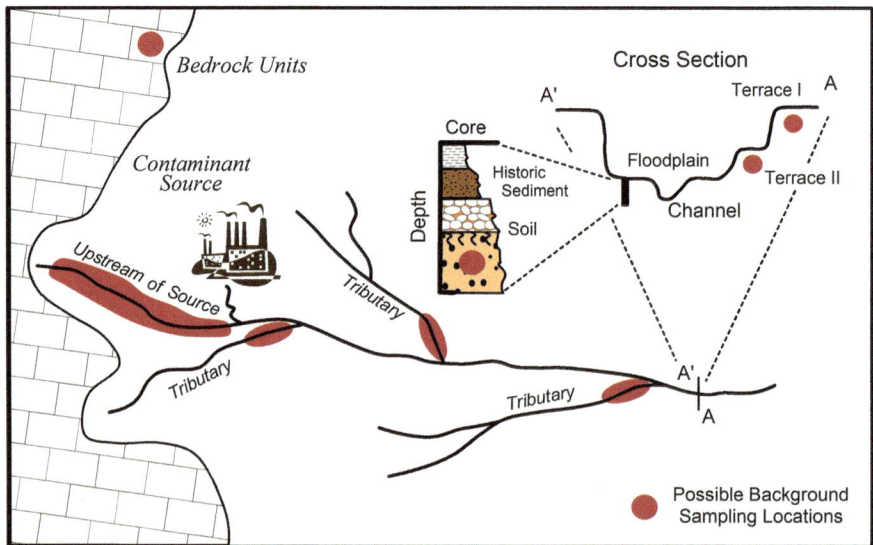

Fig. 4.7 Sites from which to determine background (geogenic) Pb isotopic ratios (modified from Miller et al. 2007)

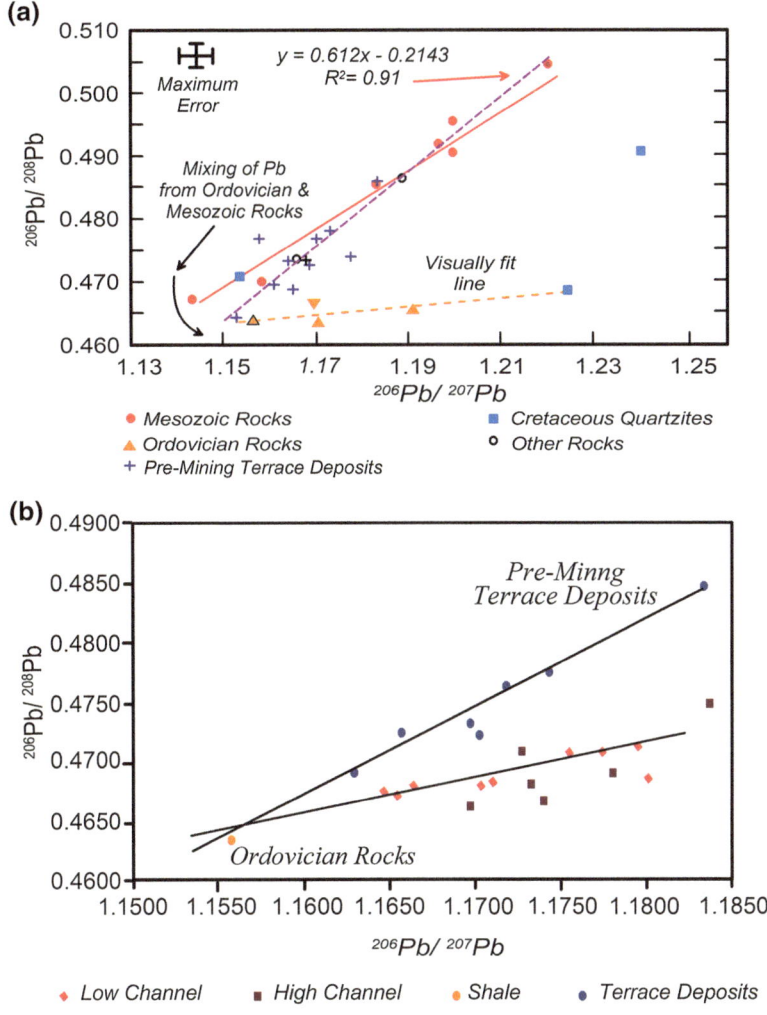

Fig. 4.8 a Pb isotopic ratios measured within Ordovician, Mesozoic and Cretaceous rocks versus those measured in pre-mining (and uncontaminated) terrace deposits (*purple dashed line*). Pre-mining terrace deposits appear to be composed of sediment from Mesozoic rocks with relatively high Pb ratios and Ordovician rocks with low ratios (from Miller et al. 2007); **b** Pb Isotopic data collected in 2000 from the modern channel bed (*red circles* and *brown squares*). High ratio end member is consistent with isotopic values measured in ore deposits mined and milled tailings from Cerro Rico, indicating Pb is derived from the mines and the underlying rocks (particularly Ordovician Rocks) (from Miller et al. 2002)

within the channel bed sediment downstream of the mills at Cerro Rico rapidly changed through time (from 2000 to 2004). The alterations were attributed, in part, to changes in the isotopic signatures of the ores that were mined and milled at Cerro Rico. In 2000, the channel bed samples downstream of the mills exhibit $^{206}Pb/^{207}Pb$

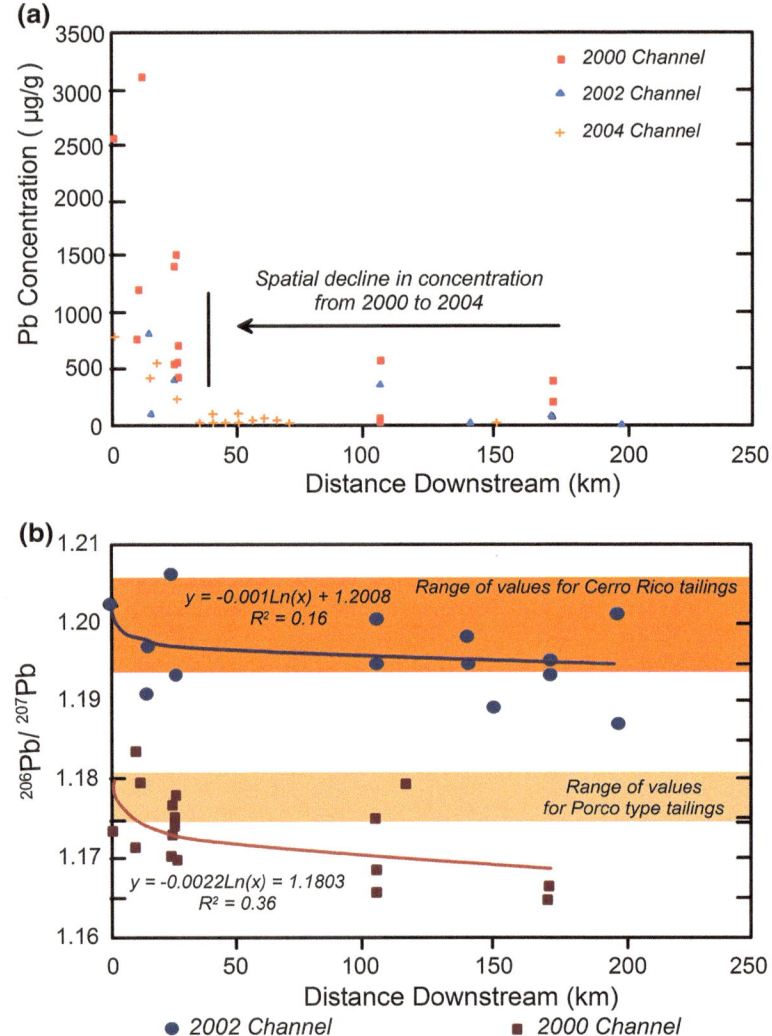

Fig. 4.9 **a** Longitudinal (downstream) changes in Pb concentrations within channel bed sediments of the Rio Pilcomayo from 2000 to 2004. **b** Comparison of $^{206}Pb/^{207}Pb$ ratios measured in channel bed sediments sampled in 2000 (*blue circles*) and 2002 (*red squares*). In 2000, $^{206}Pb/^{207}Pb$ ratios decreased downstream indicating that the relative contribution of Pb from mining operations at Cerro Rico declined as the result of mixing as 'clean' sediment from tributaries was introduced to the channel during runoff events. In 2002, precipitation and runoff events were limited. As a result, Pb ratios were generally indistinguishable from the Cerro Rico type ore deposits, suggesting that nearly all of the Pb was derived from mining operations (from Miller et al. 2007)

and $^{206}Pb/^{208}Pb$ ratios similar to what they informally referred to as Porco type ores, whereas in 2002 samples collected from along the same reach exhibited a signature comparable to Cerro Rico type ores (Fig. 4.9b). Thus, they were not only able to determine that the Pb was derived from the mines, but the ore body from which the Pb was derived.

Downstream trends in Pb isotopic ratios also reflected temporal changes in the amount of Pb derived from the mills within the channel bed sediment. In 2000, ^{206}Pb/^{207}Pb values within the lower channel bed sediments tended to decline downstream toward values observed within the Ordovician rocks, indicating a decrease in the contributions of Pb from the mills. In contrast, ^{206}Pb/^{207}Pb ratios in channel bed sediments in 2002 exhibited no downstream trend; rather, the ^{206}Pb/^{207}Pb ratios were either indistinguishable from the Cerro Rico type ore deposits, or fell between the Cerro Rico and Porco types of ore deposits (Fig. 4.9b). Changes in the isotopic ratios were coincident with a decrease in: (1) Pb concentrations at any given site along the channel; and (2) the downstream extent to which elevated Pb concentrations could be observed (Fig. 4.9a). Miller et al. (2007) argued that the changes in Pb concentration and Pb isotopic ratios reflected annual differences in system hydrology and upland/tributary sediment inputs to the river. During wet years, such as 2000, extensive runoff from the mines and mills allowed large quantities of Pb-contaminated sediment to be transported downstream over longer distances, resulting in relatively high Pb concentrations. However, the proportion of the Pb derived from the mines decreased downstream, as indicated by the isotopic ratios, because relatively large quantities of uncontaminated, but Pb-enriched fine-grained sediments eroded from upland areas were delivered to the channel via tributaries. During relatively dry years, such as 2002, the transport of Pb-contaminated sediment from the mines and mills was reduced, resulting in a decrease in Pb concentrations within the channel bed sediments. However, most of the Pb found in the channel was derived from the upstream mining operations because less sediment was eroded from the surrounding uplands and delivered to the channel. The study by Miller et al. (2007) demonstrated that the source of Pb to the channel can vary over relatively short time frames as a function of the annual variations in catchment hydrology and sediment transport processes within a catchment. This is especially true in higher energy rivers with large sediment fluxes, such as the Rio Pilcomayo.

Within larger river systems contaminated by multiple anthropogenic Pb sources, or which may be underlain by a wide range of bedrock types, the Pb isotopic signature of the alluvial sediments is likely to vary such that well-defined mixing lines do not exist on binary scatter plots (Ip et al. 2007). As a result, determining the relative contribution of Pb from the identified sources becomes much more difficult. Ayrault et al. (2012), for example, found that there were systematic trends in Pb concentration and ^{206}Pb/^{207}Pb ratios within dated cores obtained from various locations along the Seine River (Fig. 4.10a, b). However, when combined on a bivariate plot of ^{206}Pb/^{207}Pb ratios versus Pb concentration, the data formed a complex pattern. Although complex, they were able to show that the data reflected systematic changes in Pb input to the system such that two mixing lines could be defined, each representing a different time period. Between 1927 and 1968, sediments in cores B2 and M1 fell within a narrow range of values that were consistent with Pb from the Rio Tinto ore body. Pb from the Rio Tinto was thought to be used extensively for centuries in the Seine River basin, particularly after 1850, and was therefore referred to as

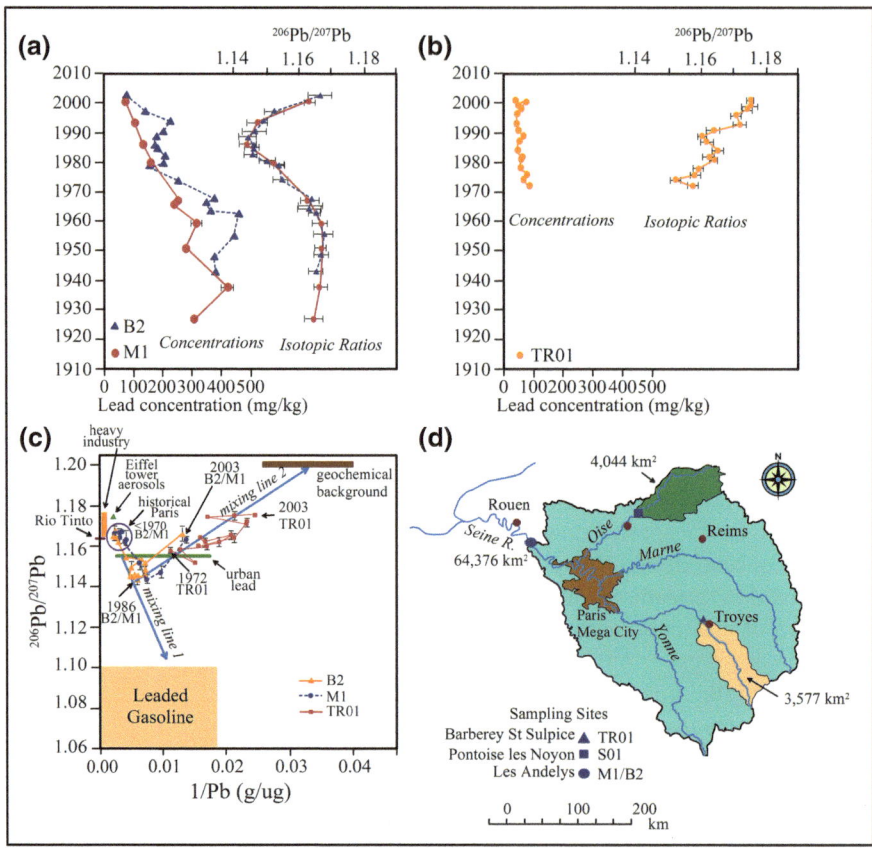

Fig. 4.10 **a–b** Variations in Pb concentration and ^{206}Pb/^{207}Pb ratios as a function of sediment age (depth) within alluvial sediment cores extracted from selected locations along the Seine River; **c** Bivariate plot showing the changes in Pb mixing through time; **d** Map showing the location of the coring/sampling sites (adapted from Ayrault et al. 2012)

the "historical" Parisian lead (Ayrault et al. 2012). From approximately 1968–1989, ^{206}Pb/^{207}Pb ratios decreased along a trend described by mixing line 1 (Fig. 4.10c). This decrease was attributed to an increase in the use of leaded gasoline (characterized by a lower isotopic ratio) and the implementation of environmental regulations that reduced other forms of Pb input to the river. As the use of leaded gasoline was phased out after 1986, Ayrault et al. (2012) argued that the Pb isotopic ratio reflected a mixture of Pb from leaded gasoline and the historical Parisian lead, which they referred to as urban lead. After about 1989, Pb concentrations decreased and ^{206}Pb/^{207}Pb signature progressively evolved (along mixing line 2) such that the Pb was composed of urban Pb and Pb from natural background sources associated with local soils (Fig. 4.10c). The fact that mixing lines could be defined allowed them to estimate the contributions of Pb from the two primary sources that were present for specific time periods. It is also important to recognize that the use of dated cores allowed an historical understanding of Pb use and provenance in the catchment to be determined.

Bird et al. (2010a, b) overcame the multiple source problem by using a two-tiered approach to assess Pb provenance within the River Maritsa catchment of Bulgaria and Turkey. A binary isotopic approach was used within smaller tributary basins to assess the contributions of Pb from mining operations to the channel. However, highly variable Pb isotopic values inhibited the use of this binary method within lower catchment areas where sediment was derived from a wider range of geological units. Thus, sediment/Pb provenance was determined for the lower catchment areas using multi-element geochemical fingerprinting methods similar to that described in Chap. 2. Bird et al. (2010a, b) also applied a similar approach to the lower Danube Catchment in Eastern Europe. The fingerprinting approach has also been applied directly to Pb isotopic data to assess Pb provenance where multiple Pb sources exist (Miller et al. 2007). A rather different graphical approach to quantifying Pb contributions has occasionally been used where three end-member sources can be identified on the basis of Pb concentration and a Pb isotopic ratio (e.g., Lima et al. 2005; N'guessan et al. 2009). The approach is based on the graphical plotting of linear mixing lines between the three end-member sources on the basis of sample concentrations and their associated isotopic ratios forming a ternary diagram. Once the ternary diagram has been created, the relative contributions of Pb from the three delineated sources then can be estimated from the plot.

4.4 Summary and Management Implications

Sr and Nd isotopic systems have been extensively used as tracers to address such problems as weathering rates and intensities, the potential factors influencing temporal variations in seawater chemistry, and the source of particulates in dust and sedimentary rocks. Their application to historic and contemporary riverine environments, particularly at small spatial scales, is more restricted, as is their use to source contaminated particles. However, their proven potential to source sedimentary particles and recent studies suggest that both Sr and Nd isotopes may be used to differentiate between various types of anthropogenic pollutants indicates that their application to riverine contaminant issues will prove effective. Moreover, their application to contaminated rivers will be able to make use of an extensive body of literature on the topic that has been generated for the analysis of other geological systems. Their utilization will likely be driven by the inability of a single isotopic system (e.g., Pb) to distinguish between the known contaminant sources. Thus, we suspect that Sr and Nd isotopes will be of most value when used in combination with Pb isotopes and/or other elemental tracers.

Of these three isotope systems discussed herein, Pb has been the most extensively utilized as a tracer of anthropogenic contaminants. Recent studies have demonstrated that Pb isotopes can be effectively applied to riverine sediments to determine the relative contributions of material from a suspected contaminant source and the dispersal rates and transport pathways from the source(s). Moreover, Pb isotopes have been effectively applied to systematically accumulated and dated deposits to assess changes in contaminant flux to a river through time. Pb isotopes, then, represent a powerful tool for determining the source of Pb contaminated sediment when the source materials can be distinguished on the basis of Pb isotopic ratios. The use of Pb isotopes will be more effective within smaller catchments with a limited number of potential Pb sources than within larger basins characterized by a wider range of rock types with varying Pb isotopic ratios. Limitations of the use of Pb isotopes are related in part to the inability to effectively distinguish between the Pb of differing industrial contaminants, and the lack of well-defined protocols for quantitatively estimating the uncertainties associated with the estimation of Pb contributions from the delineated sources. The quantification of uncertainties is particularly needed for environmental forensic investigations. Unfortunately, the signature of end-member Pb sources has been based in most studies on relatively simple statistics of the source data, such as the mean. As a result, variations in isotopic ratios inherent in the source materials (or the sampled river sediments) are not quantitatively considered in the estimation of Pb contributions, making it difficult to assess the uncertainties involved in the analysis. More attention will need to be given to the use of rigorous statistical methods to (1) characterize the variation in Pb isotopic ratios within the source and river sediments, and (2) utilize an understanding of the variations in the source ascription process before Pb isotopic methods can be extensively used in forensic analysis. Method development may benefit from recent advances that have been made in the geochemical fingerprinting of non-point source pollutants described in Chap. 2.

Future analyses using Pb isotopes are likely to focus more on the cycling of Pb within the riverine environment, including its dispersal from a point source in both the particulate and dissolved forms, its sorption and desorption from sediments, and its accumulation in biota. Particular attention is likely to be paid to the use of Pb isotopes to quantitatively determine the provenance of Pb in aquatic biota. While existing studies suggest that Pb isotopes hold considerable promise in determining the source of Pb in biota, the analysis is unlikely to be as straightforward as might be expected.

Miller et al. (2005), for example, used Pb isotopes to examine the accumulation of Pb associated with orchard soils contaminated by lead arsenate in Rainbow Trout (*Oncorhynchus mykiss*) in the Richland Creek basin of Western North Carolina. They found that Pb concentrations and Pb isotopic ratios varied between muscle, liver, and bone (Fig. 4.11). Although differences in Pb concentration were expected, differences in Pb isotopic abundances were not because previous studies had suggested that biological processes do not significantly fractionate Pb. Miller et al. (2005) attributed the observed difference in isotopic ratios to the temporary exposure of the trout to contaminated waters, and to subsequent differences in the ability of the tissues to excrete Pb. In fact, laboratory studies demonstrated that when rainbow trout were exposed to high levels of dissolved Pb with a distinct isotopic signature, both the bone and liver rapidly acquired Pb from the water and took on its isotopic ratios. Following exposure the bone retained the Pb and the acquired Pb isotopic ratios for a period of months. In contrast, the liver excreted approximately 50 % of the

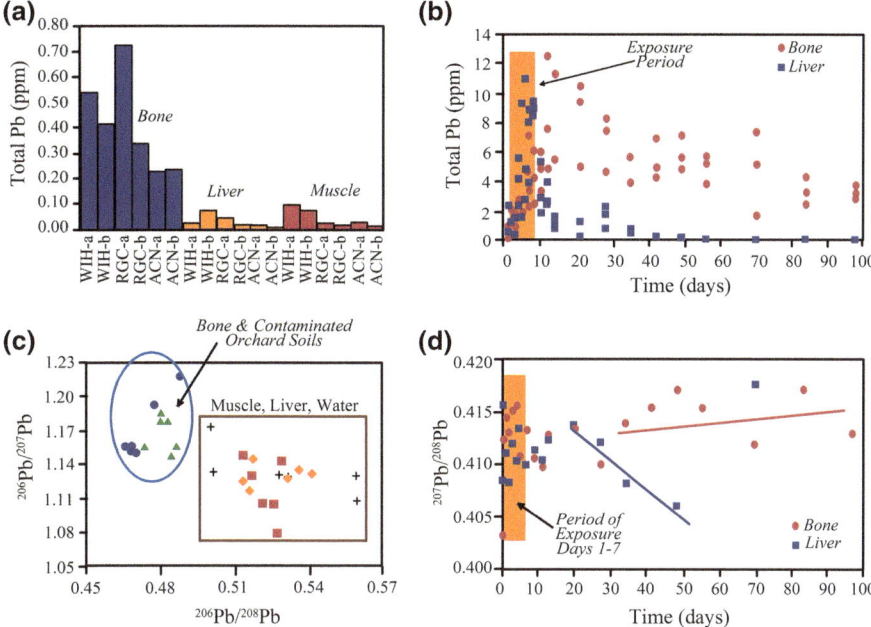

Fig. 4.11 a Lead concentrations measured in bone, liver and muscle within rainbow trout from Richland Creek, western North Carolina; **b** Temporal changes in Pb concentrations in fish subject to a seven-day dose of lead nitrate. Concentrations in bone and liver decrease as different rates; **c** Pb isotopic ratios in fish from Richland Creek; **d** Temporal changes in $^{207}Pb/^{208}Pb$ ratios of laboratory trout following exposure to a Pb source. No statistically significant change occurred in bone, whereas a systematic change was observed in liver tissues. The latter change correlated with rapid decreases in Pb concentrations (adapted from Miller et al. 2005)

accumulated Pb within a few days, while nearly all of the Pb was excreted within a few weeks. As the Pb was excreted, the isotopic signature of the liver and muscle changed toward that of the 'clean' water (Fig. 4.11). Additional studies are needed to determine if such phenomena affect other biota in other environments. Nonetheless, the Richland Creek study suggests that the Pb isotopic analysis of trout may not only provide insights into whether biota is exposed to a particular Pb source, but the length of exposure. In the case of rainbow trout, for example, it seems possible that the isotopic composition of fish bone can be used as an indicator of long-term Pb exposure whereas the composition of the liver and muscle will provide insights into their shorter-term exposure to Pb (Miller et al. 2005).

References

Abouchami W, Galer SJ, Koschinsky A (1999) Pb and Nd isotopes in NE Atlantic Fe-Mn crusts: proxies for trace metal paleosources and paleocean circulation. Geochim Cosmochim Acta 63:1489–1505

Allègre CJ (2008) Isotope geology. Cambridge University Press, New York

Ault WU, Senechal RG, Erlebach WE (1970) Isotopic composition as a natural tracer of lead in the environment. Environ Sci Technol 4:305–313

Awwiller DN (1994) Geochronology and mass transfer in Gulf Coast mudrocks (South-Central Texas, USA): Rb-Sr, Sm-Nd and REE systematics. Chem Geol 116:61–84

Awwiller DN, Mack LE (1989) Diagenetic resetting of Sm-Nd isotope systematics in Wilcox group sandstones and shales, San Marcos Arch, South-Central Texas. Gulf Coast Association Geological Society Transactions, New Orleans

Ayrault S, Roy-Barman M, Le Cloarec MF, Rianti Priadi C, Bonté P, Göpel C (2012) Lead contamination of the Seine River, France: geochemical implications of a historical perspective. Chemosphere 87:902–910

Balcaen L, Moens L, Vanhaecketitle F (2010) Determination of isotope ratios of metals (and metalloids) by means of inductively coupled plasma-mass spectrometry for provenancing purposes A review. Spectrochim Acta B 65:769–786

Banner JL (2004) Radiogenic isotopes: systematics and applications to earth surface processes and chemical stratigraphy. Earth Sci Rev 65:141–194

Bindler R, Klarqvist M, Klaminder J, Forster J (2004) Does within-bog spatial variability of mercury and lead constrain reconstructions of absolute deposition rates from single peat records? The example of Store Mosse, Sweden. Glob Biogeochem Cycles 18:1–11

Bird G (2011) Provenancing anthropogenic Pb within the fluvial environment: developments and challenges in the use of Pb isotopes. Environ Int 37:802–819

Bird G, Brewer PA, Macklin MG, Nikolova M, Kotsev T, Mollov M, Swain C (2010a) Quantifying sediment-associated metal dispersal using Pb isotopes: application of binary and multivariate mixing models at the catchment-scale. Environ Pollut 158:2158–2169

Bird G, Brewer PA, Macklin MG, Nikolova M, Kotsev T, Mollov M, Swain C (2010b) Pb isotope evidence for contaminant-metal dispersal in an international river system: the lower Danube catchment, Eastern Europe. Appl Geochem 25:1070–1084

Biscaye PE, Grousset FE, Revel M, Van der Gaast S, Zielinski GA, Vaars A, Kukla G (1997) Asian provenance of glacial dust—stage 2. in the Greenland Ice Sheet Project 2 Ice Core Summit, Greenland. J Geophys Res 102:26765–26781

Basile I, Grousset FE, Revel M, Petit JR, Biscaye PE, Barkov NI (1997) Patagonian origin of glacial dust deposited in East Antarctica—Vostok and Dome C. during glacial stages 2, 4 and 6. Earth Planet Sci Lett 146:573–589

Basile I, Petit JR, Touron S, Grousset FE, Barkov N (2001) Volcanic ash layers in Vostok ice–core: source identification and atmospheric implications. J Geophys Res 106:31915–31928

Bellis DJ, Satake K, McLeod CW (2004) A comparison of lead isotope ratios in the bark pockets and annual rings of two beech trees collected in Derbyshire and South Yorkshire, UK. Sci Total Environ 321:105–113

Bi X, Feng X, Yang Y, Li X, Shin GPY, Li F, Qiu G, Li G, Liu T, Fu Z (2009) Allocation and source attribution of lead and cadmium in maize (Zea mays L.) impacted by smelting emissions. Environ Pollut 157:834–839

Calanchi N, Dinelli E, Lucchini F, Mordenti A (1996) Chemostratigraphy of late Quaternary sediments from Lake Albano and central Adriatic Sea cores (PALICLAS Project). Memorie-Istituto Italiano di Idrobiologia 55:247–264

Cheng H, Hu Y (2010) Lead (Pb) isotopic fingerprinting and its applications in lead pollution studies in China: a review. Environ Pollut 158:1134–1146

Chiaradia M, Cupelin F (2000) Behaviour of airborne lead and temporal variations of its source effects in Geneva (Switzerland): comparison of anthropogenic versus natural processes. Atmos Environ 34:959–971

Chiaradia M, Chenhall BE, Depers AM, Gulson BL, Jones BG (1997) Identification of historical lead sources in roof dusts and recent lake sediments from an industrialized area: indications from lead isotopes. Sci Total Environ 205:107–128

Church SE, Unruh DM, Fey DL, Sole TC (2004) Trace elements and lead isotopes in streambed sediment in streams affected by historical mining. US Geological Survey professional paper no 1652:279–336

Chow TJ, Johnstone MS (1965) Lead Isotopes in gasoline and aerosols of Los Angeles Basin, California. Science 147:502–503

Cloy JM, Farmer JG, Graham MC, MacKenzie AB, Cook GT (2008) Historical records of atmospheric Pb deposition in four Scottish ombrotrophic peat bogs: an isotopic comparison with other records from western Europe and Greenland. Glob Biogeochem Cycles 22:GB2016

Dasch EJ (1969) Strontium isotopes in weathering profiles, deep-sea sediments, and sedimentary rocks. Geochim Cosmochim Acta 33:1521–1552

DePaolo DJ (1981) A neodymium and strontium isotopic study of the Mesozoic calcalkalic granitic batholiths of the Sierra-Nevada and Peninsular Ranges, California. J Geophys Res 86:10470–10488

Dunham KC (1990) Geology of the Northern Pennine orefield, Tyne to Stainmore. Econ Mem Brit Geol Surv 1

Elbaz-Poulichet F, Holliger P, Martin JM, Petit D (1986) Stable lead isotopes ratios in major french rivers and estuaries. Sci Total Environ 54:61–76

Erel Y, Veron A, Halicz L (1997) Tracing the transport of anthropogenic lead in the atmosphere and in soils using isotopic ratios. Geochim Cosmochim Acta 61:4495–4505

Foster IDL, Lees JA, Jones AR, Chapman AS, Turner SE, Hodgkinson R (2002) The possible role of agricultural land drains in sediment delivery to a small reservoir, Worcestershire, UK: a multiparameter fingerprint study. In: The structure, function and management of fluvial sedimentary systems, Alice Springs, AU. Int Assoc Hydrol Sci, Alice Springs, Australia, No 276:433–442

Goldstein SL, O'Nions RK, Hamilton PJ (1984) A Sm-Nd isotopic study of atmospheric dusts and particulates from major river systems. Earth Planet Sci Lett 70:221–236

Goldstein SJ, Jacobsen SB (1988) Nd and Sr isotopic systematics of river suspended material: implications for crustal evolution. Earth Planet Sci Lett 87:249–265

Grousett FE, Biscaye PE (2005) Tracing dust sources and transport patterns using Sr, Nd, and Pb isotopes. Chem Geol 222:149–167

Grousset FE, Biscaye PE, Revel M, Petit JR, Pye K, Joussaume S, Jouzel J (1992) Antarctic ice core dusts at 18 ky B.P.: isotopic constraints on origins and atmospheric circulation. Earth Planet Sci Lett 111:175–182

Guéguen F, Stille P, Perrone T, Chabaux F (2012) Atmospheric pollution in an urban environment by tree bark biomonitoring—Part II: Sr, Nd, Pb isotopic tracing. Chemosphere 86:641–647

Gulson BL, Tiller KG, Mizon KJ, Merry RH (1981) Use of lead isotopes in soils to identify the source of lead contamination near Adelaide, South Australia. Environ Sci Technol 15:691–696

Gulson BL, Mizon KJ, Korsch MJ, Howarth D (1996) Importance of monitoring family members in establishing sources and pathways of lead in blood. Sci Total Environ 188:173–182

Gulson BL, Howarthl D, Mizon KJ, Law AJ, Korsch MJ, Davis JJ (1994) Source of lead in humans from Broken Hill mining community. Environ Geochem Health 16:19–25

Hansmann W, Köppel V (2000) Lead-isotopes as tracers of pollutants in soils. Chem Geol 171:123–144

Hoefs J (2009) Stable Isotope Geochemistry. Springer Verlag, New York

Hopper JF, Ross HB, Sturges WT, Barrie LA (1991) Regional source discrimination of atmospheric aerosols in Europe using the isotopic composition of lead. Tellus B 43:45–60

Hudson-Edwards KA, Macklin MG, Miller JR, Lechler PJ (2001) Sources, distribution and storage of heavy metals in the Rio Pilcomayo, Bolivia. J Geochem Explor 72:229–250

Ip CCM, Li XD, Zhang G, Wai OWH, Li YS (2007) Trace metal distribution in sediments of the Pearl River Estuary and the surrounding coastal area, South China. Environ Pollut 147:311–323

Jones CE, Halliday AN, Rea DK, Owen RM (1994) Neodymium isotopic variations in North Pacific modern silicate sediment and the insignificance of detrital REE contributions to seawater. Earth Planet Sci Lett 127:55–66

Kamenov GD, Brenner M, Tucker JL (2009) Anthropogenic versus natural control on trace element and Sr-Nd-Pb isotope stratigraphy in peat sediments of southeast Florida (USA), ~1500 to present. Geochim Cosmochim Acta 73:3549–3567

Keinonen M (1992) The isotopic composition of lead in man and the environment in Finland 19661987: isotope ratios of lead as indicators of pollutant source. Sci Total Environ 113:251–268

Kristensen LJ, Taylor MP, Odigie K, Hibdon SA, Flegal AR (2014) Lead isotopic compositions of ash sourced from Australian bushfires. Environ Pollut 190:159–165

Komárek M, Ettler V, Chrastný V, Mihaljeviç M (2008) Lead isotopes in environmental sciences: a review. Environ Int 34:562–577

Kurkjian R, Dunlap C, Flegal AR (2004) Long-range downstream effects of urban runoff and acid mine drainage in the Debed River, Armenia: insights from lead isotope modeling. Appl Geochem 19:1567–1580

Lahd Geagea M, Stille P, Millet M, Perrone T (2007) REE characteristics and Pb, Sr and Nd isotopic compositions of steel plant emissions. Sci Total Environ 373:404–419

Lahd Geagea M, Stille P, Gauthier-Lafaye F, Millet M (2008a) Tracing of industrial aerosol sources in an urban environment using Pb, Sr, and Nd isotopes. Environ Sci Technol 42:692698

Lahd Geagea M, Stille P, Gauthier-Lafaye F, Perrone T, Aubert D (2008b) Baseline determination of the atmospheric Pb, Sr and Nd isotopic compositions in the Rhine Valley, Vosges Mountains (France) and the Central Swiss Alps. Appl Geochem 23:1703–1714

Laidlaw MAS, Zahran S, Pingatore N, Clague J, Devlin G, Taylor MP (2014) Identifying and fingerprinting temporal lead sources in domestic homes. Environ Pollut 184:238246

Lima AL, Bergquist BA, Boyle EA, Reuer MK, Dudas FO, Reddy CM, Eglinton TI (2005) High resolution Pb isotope record reveals a stratigraphic marker for the Northeast USA. Geochim Cosmochim Acta 69:1813–1824

Mackay AK, Taylor MP, Munksgaard NC, Hudson-Edwards KA, Burn-Nunes L (2013) Identification of environmental lead sources, pathways and forms in a mining and smelting town: Mount Isa, Australia. Environ Pollut 180:304–311

Manton WI (1977) Sources of lead in blood: identification by stable isotopes. Arch Environ Health 32:149–159

Marcantonio F, Flowers GC, Templin N (2000) Lead contamination in a wetland watershed: isotopes as fingerprints of pollution. Environ Geol 39:1070–1076

McCulloch MT, Wasserburg GJ (1978) Sm-Nd and Rb-Sr chronology of continental crust formation. Science 200:1003–1011

McGill RAR, Pearce JM, Fortey NJ, Watt J, Ault L, Parrish RR (2003) Contaminant source apportionment by Pimms lead isotope analysis and Sem-Image analysis. Environ Geochem Health 25:25–32

McLennan SM (1989) Rare elements in sedimentary rocks: influence of provenance and sedimentary processes. In: Lipin BR, McKay GA (eds) Geochemistry and mineralogy of rare earth elements. Rev Miner 21:169–200

McLennan SM, Taylor SR, McCulloch MT, Maynard JB (1990) Geochemical and Nd-Sr isotopic compositions of deep-sea turbidites: Crustal evolution and plate tectonic associations. Geochim Cosmochim Acta 54:2015–2050

McLennan SM, Hemming S, McDaniel DK, Hanson GN (1993) Geochemical approaches to sedimentation, provenance and tectonics. In: Johnsson MJ, Basu A (eds) Processes controlling the composition of clastic sediments. GSA Spec Paper 284:21–40

Mihaljeviç M, Ettler V, Šebek O, Strnad L, Chrastný V (2006) Lead isotopic signatures of wine and vineyard soilstracers of lead origin. J Geochem Explor 88:130–133

Miller JR, Orbock Miller SM (2007) Contaminated rivers: a geomorphological-geochemical approach to site assessment and remediation. Springer, Berlin

Miller JR, Lechler PJ, Hudson-Edwards KA, Macklin MG (2002) Lead isotopic fingerprinting of heavy metal contamination, Rio Pilcomayo basin, Bolivia. Geochem Explor Environ Anal 2:225–233

Miller J, Lord M, Yurkovich S, Mackin G, Kolenbrander L (2005) Historical trends in sedimentation rates and sediment provenance, Fairfield Lake, Western North Carolina. JAWRA 41:1053–1075

Monna F, Lancelot J, Croudace IW, Cundy AB, Lewis JT (1997) Pb isotopic composition of airborne particulate material from France and the Southern United Kingdom? Implications for Pb pollution sources in Urban areas. Environ Sci Technol 31:2277–2286

Moon S, Huh Y, Qin J et al (2007) Chemical weathering in the Hong (Red) River basin: rates of silicate weathering and their controlling factors. Geochim Cosmochim Acta 71:1411–1430

Nelson BK, DePaolo DJ (1988) Application of Sm-Nd and Rb-Sr isotope systematics to studies of provenance and basin analysis. J Sed Petrol 58:348–357

N'guessan YM, Probst JL, Bur T, Probst A (2009) Trace elements in stream bed sediments from agricultural catchments (Gascogne region, S-W France): Where do they come from? Sci Total Environ 407:2939–2952

Padoan M, Garzanti E, Harlavan Y, Villa IM (2011) Tracing Nile sediment sources by Sr and Nd isotope signatures (Uganda, Ethiopia, Sudan). Geochimi Cosmochimi Acta 75:36273644

Petit D, Mennessier JP, Lambertstitle L (1984) Stable lead isotopes in pond sediments as tracer of past and present atmospheric lead pollution in Belgium. Atmos Environ 18:1189–1193

Potot C, Féraud G, Schärer U, Barats A, Durrieu G, Le Poupon C, Travi Y, Simler R (2012) Groundwater and river baseline quality using major, trace elements, organic carbon and SrPbO isotopes in a Mediterranean catchment: the case of the Lower Var Valley (South-Eastern France). J Hydrol 89:472–473

Rai SK (2007) Temporal variations in Sr and $^{87}Sr/^{86}Sr$ of the Brahmaputra: implications for annual fluxes and tracking flash floods through chemical and isotope composition. Geochem Geophys Geosyst 8:Q08008

Reeder M, Rothwell RG, Stow DAV, Kahler G, Kenyon NH (1998) Turbidite flux, architecture and chemostratigraphy of the Herodotus Basin, Levantine Basin, SE Mediterranean Sea. In: Stoker MS, Evans D, Cramp A (eds) Geological processes on continental margin: sedimentation, mass-wasting and stability. Geol Soc Spec Publ 129:19–42

Reimann C, Flem B, Fabian K, Birke M, Ladenberger A, Négrel P, Demetriades A, Hoogewerff J (2012) Lead and lead isotopes in agricultural soils of Europe the continental perspective. Appl Geochem 27:532–542

Rosenbauer RJ, Foxgrover AC, Hein JR, Swarzenski PW (2013) A Sr-Nd isotopic study of sand-sized sediment provenance and transport for the San Francisco Bay coastal system. Mar Geol 345:143–153

Rosman KJR, Chisholm W, Boutron CF, Candelone JP, Görlach U (1993a) Isotopic evidence for the source of lead in Greenland snows since the late 1960s. Nature 362:333–335

Rosman KJR, Chisholm W, Boutron CF, Candelone JP, Patterson CC (1993b) Anthropogenic lead isotopes in Antarctica. Geophys Res Lett 21:2669–2672

Rosman KJR, Chisholm W, Hong S, Candelone JP, Boutron CF (1997) Lead from Carthaginian and Roman Spanish mines isotopically identified in Greenland Ice dated from 600 B.C. to 300 A.D. Environ Sci Technol 31:3413–3416

Sangster DF, Outridge PM, Davis WJ (2000) Stable lead isotope characteristics of lead ore deposits of environmental significance. Environ Rev 8:115–147

Shepherd TJ, Chenery SRN, Pashley V, Lord RA, Ander LE, Breward N, Hobbs SF, Matthew Horstwood M, Klinck BA, Worrall F (2009) Regional lead isotope study of a polluted river catchment: River Wear, Northern England, UK. Sci Total Environ 407:4882–4893

Shirahata H, Elias RW, Patterson CC, Koide M (1980) Chronological variations in concentrations and isotopic compositions of anthropogenic atmospheric lead in sediments of a remote subalpine pond. Geochim Cosmochim Acta 44:149–162

Shotyk W, Weiss D, Appleby PG, Cheburkin AK, Frei R, Gloor M, Kramers JD, Reese S, Van der Knaap WO (1998) History of atmospheric lead deposition since 12,370 14C yr BP from a Peat Bog, Jura Mountains, Switzerland. Science 281:1635–1640

Shotyk W, Zheng J, Krachler M, Zdanowicz C, Koerner R, Fisher D (2005) Predominance of industrial Pb in recent snow (19942004) and ice (18421996) from Devon Island, Arctic Canada. Geophys Res Lett 32

Sondergaard J, Asmund G, Johansen P, Elberling B (2010) Pb isotopes as tracers of mining-related Pb in lichens, seaweed and mussels near a former Pb-Zn mine in West Greenland. Environ Pollut 158:1319–1326

Soto-Jiménez MF, Páez-Osuna F, Scelfo G, Hibdon S, Franks R, Aggarawl J, Flegal AR (2008) Lead pollution in subtropical ecosystems on the SE Gulf of California Coast: a study of concentrations and isotopic composition. Mar Environ Res 66:451–458

Soto-Jiménez MF, Flegal AR (2009) Origin of lead in the Gulf of California Ecoregion using stable isotope analysis. J Geochem Explor 101:209–217

Steinmann M, Stille P (1997) Rare earth element behavior and Pb, Sr, Nd isotope systematics in a heavy metal contaminated soil. Appl Geochem 12:607–623

Tripathy GR, Goswami V, Singh SK et al (2010) Temporal variations in Sr and ^{87}Sr/^{86}Sr of the Ganga headwaters: estimate of dissolved Sr flux to the mainstream. Hydrol Process 24:1159–1171

Tripathy GR, Singh SK, Krishnaswami S (2011) Sr and Nd isotopes as tracers of chemical and physical erosion. In: Baskaran M (ed) Handbook of environmental isotope geochemistry. Advances in Isotope Geochemistry. Springer, Berlin, pp 521–552

Veysseyre AM, Bollhöfer AF, Rosman KJR, Ferrari CP, Boutron CF (2001) Tracing the origin of pollution in French alpine snow and aerosols using lead isotopic ratios. Environ Sci Technol 35:4463–4469

Voss BM, Peucker-Ehrenbrink B, Eglinton TI, Fiske G, Aleck Wang Z, Hoering KA, Montluon DB, LeCroy C, Pal S, March S, Gillies SL, Janmaat A, Bennett M, Downey B, Fanslau J, Fraser H, Macklam-Harron G, Martinec M, Wiebe B (2014) Tracing river chemistry in space and time: dissolved inorganic constituents of the Fraser River, Canada. Geochim Cosmochim Acta 124:283–308

Walter HJ, Hegner E, Diekmann B et al (2000) Provenance and transport of terrigenous sediment in the south Atlantic Ocean and their relations to glacial and interglacial cycles: Nd and Sr isotopic evidence. Geochim Cosmochim Acta 64:3813–3827

Weiss D, Shotyk W, Appleby PG, Kramers JD, Cheburkin AK (1999) Atmospheric Pb deposition since the industrial revolution recorded by five Swiss peat profiles: enrichment factors, fluxes, isotopic composition, and sources. Environ Sci Technol 33:1340–1352

Weldeab S, Emeis KC, Hemleben C, Siebel W (2002) Provenance of lithogenic surface sediments and pathways of riverine suspended matter in the Eastern Mediterranean Sea: evidence from ^{143}Nd/^{144}Nd and ^{87}Sr/^{86}Sr ratios. Chem Geol 186:139–149

Wei G, Liu Y, Ma J, Xie L, Chen J, Deng W, Tang S (2012) Nd, Sr isotopes and elemental geochemistry of surface sediments from the South China Sea: implications for provenance tracing. Mar Geol 319:21–34

Winter BL, Johnson CM, Clark DL (1997) Strontium, neodymium, and lead isotope variations of authigenic and silicate sediment components from the Late Cenozoic Arctic Ocean: implications for sediment provenance and the source of trace metals in seawater. Geochim Cosmochim Acta 61:4181–4200

Yeager KM, Santschi PH, Herbert BE (2005) Suspended sources and tributary effects in the lower reaches of a coastal plain stream as indicated by radionuclides, Loca Bayou, Texas. Environ Geol 47:382–395

Chapter 5
Stable 'Non-Traditional' Isotopes

Abstract Recent advances in analytical capabilities, particularly the MC-ICP-MS, have allowed for a precise determination of a wide range of stable isotopes in geological and biological materials that could not be assessed prior to the 1990s. As a result, research into the use of these 'non-traditional' isotopes (or 'non-CHONS') as tracers of both elemental sources and biogeochemical processes has been increasing at an exponential rate. While their utilization as a tracer of contaminated sediments in the near surface environment is often complicated by multiple physical and biological fractionation processes, there is increasing evidence to suggest that they may be effectively used as tracers in aquatic environments. In this chapter, we examine the potential use of four stable metal isotopes (Zn, Cd, Cu, and Hg) that appear on the basis of the limited studies conducted to date to have the potential to track sediment-associated trace metals in rivers.

Keywords Zn isotopes · Cd isotopes · Cu isotopes · Hg isotopes · Non-traditional sediment tracers

5.1 Introduction

In addition to the radiogenic isotopes examined in the previous chapter, significant attention has been given to various stable isotopes for elucidating Earth surface processes, particularly the isotopes of H, C, N, O, S, and Si. The isotopes of C and O have been particularly well investigated and utilized; in fact, more than 10,000 papers, abstracts, and thesis have been published on C and O isotopic variations alone since the first measurements were made in the 1930s, the majority to investigate surface and near-surface biogeochemical processes (Baskaran 2011). Study of the remaining stable isotopes (e.g., those of the transition and post-transition elements) was hampered before the 1990s by an inability to precisely measure their abundance in geological and biological materials. However, recent advances in analytical instrumentation (including the development of the MC-ICP-MS) have resulted in a dramatic increase in their potential application as environmental tracers (see Baskaran 2011 for a brief history of instrumental development). These isotopes are now frequently referred to as the 'non-traditional' isotopes.

© The Author(s) 2015
J.R. Miller et al., *Application of Geochemical Tracers to Fluvial Sediment*,
SpringerBriefs in Earth Sciences, DOI 10.1007/978-3-319-13221-1_5

The term 'non-traditional' was originally coined to differentiate the investigation of these alternative isotopes from the much more intensively studied and applied isotopes of C, H, O, N, and S. Bullen (2013) objected to the continued use of 'non-traditional' to describe work on these lesser studied isotopes, arguing that recent research has rendered the term obsolete. Moreover, he argued that the term posed "more of a roadblock than a proper description of the field" and suggested that use may lead to the risk of "having the science relegated to a niche market that is difficult for other scientists to access". He went on to propose the term of 'non-CHONS' as a replacement for non-traditional to distinguish these isotopes from the more extensively studied isotopes of C, H, O, N, and S. While we have elected to use the term non-CHONS here, it seems highly unlikely that their study will be relegated to the back corners of isotope geochemistry anytime soon. On the contrary, their study appears to be accelerating at a nearly exponential rate as they are being used to address a wide range of issues, such as the paleochemistry of the oceans, metal transfer processes in both physical and biological systems, and the identification of the source and cycling of metal and metalloid contaminants (Bullen and Walczyk 2009). With regards to the latter, interest in the isotopes stems from the fact that a number of metals (Cd, Cr, Cu, Hg, Ni, Se, Ag, and Zn) are considered by the USEPA and other regulatory bodies as priority pollutants. Thus, if their isotopes can be used as tracers, then it may be possible to directly determine their source, dispersal patterns/rates, and/or cycling processes in near-surface systems.

At the present time, use of the non-CHONS is still in the developmental stages; while some notably successes have demonstrated their potential use as environmental tracers, their application is much more difficult than, for example, the more traditional radiogenic isotopes of Pb, Nd, or Sr. The primary difference rests on the magnitude of isotopic fractionation by physical and biological processes. In the case of Pb, for example, fractionation by industrial or biological processes is negligible and the Pb isotopic composition of an anthropogenic substance depends upon the ore deposits from which the Pb was derived. Since the Pb isotopic signature of metallic ores generally differs from most other (geogenic) rocks and minerals, the measured Pb isotopic ratios within alluvial sediments results exclusively from the mixing of particles with different isotopic abundances, and anthropogenic materials can often be isotopically fingerprinted to determine the source of Pb in the river system. In contrast, variations in the isotopic abundances of non-CHONS tend to be more limited within natural geological materials (e.g., rocks, minerals, and ore deposits). However, small but measurable variations in the isotopic composition of a material derived from the original rocks, minerals, ores, etc., may occur as a result of physical and biological isotopic fractionation processes (Table 5.1) (Wombacher et al. 2004; Cloquet et al. 2006; Shiel et al. 2010; Rehkämper et al. 2011). These new isotopic abundances, which may, for example, be linked to a specific industrial process, may then be utilized as a geochemical tracer to determine a contaminant source (Bullen 2011; Rehkämper et al. 2011). Alternatively, low temperature fractionation of elements which participate in redox reactions may provide insights into biogeochemical cycling processes.

In the following sections, we turn our attention to four stable metal isotopes (Cd, Cu, Hg, and Zn) that appear on the basis of the limited studies that have been con-

Table 5.1 Selected isotope characteristics and examples of use as a geochemical tracer (modified from Miller 2013)

Element	Stable isotopes*	Frequently used isotopic ratios[1]	Primary fractionation processes	Comments	Selected references relevant to riverine environments
Cd	^{106}Cd, ^{108}Cd, ^{110}Cd, ^{111}Cd, ^{112}Cd, ^{113}Cd, ^{114}Cd, ^{116}Cd	$^{114}Cd/^{110}Cd$	Evaporation and condensation, including that associated with ore processing; biological processes	Possible use to assess nutrient uptake and translocation in microorganisms and plants; may be used to trace source, but not adequately tested	Gao et al. (2008), Cloquet et al. (2006), Weiss et al. (2008), Shiel et al. (2010, 2012)
Cu	^{63}Cu, ^{65}Cu	$^{65}Cu/^{63}Cu$	Redox reaction; biogenic accumulation; solution speciation	May be used to assess redox processes during biogeochemical cycling and to trace source of Cu, but not adequately tested	Weiss et al. (2008), Bird (2011), Kimball et al. (2009), Zhu et al. (2000), Petit et al. (2008)
Hg	^{196}Hg, ^{198}Hg, ^{199}Hg, ^{200}Hg, ^{201}Hg, ^{202}Hg, ^{204}Hg	$^{202}Hg/^{198}Hg$	MDFRedox reactions, biomethylation, evaporation, condensation; absorption; MIFphotoreduction, biological transformations	May be used to assess redox processes during biogeochemical cycling of elements	Mil-Homens et al. (2013), Bergquist and Blum (2009), Yin et al. (2010), Sonke et al. (2010)
Zn	^{64}Zn, ^{66}Zn, ^{67}Zn, ^{68}Zn, ^{70}Zn	$^{66}Zn/^{64}Zn$	Evaporation and condensation, including that associated with ore processing; biogenic uptake; diffusion, adsorption/ complexation on organics and Fe oxides	Use appears promising for tracing Zn sources from industrial processes, particularly from ore processing; may be used to assess Zn cycling in the environment	Bird (2011), Sonke et al. (2010), Shiel et al. (2012), Bentahila et al. (2008), Petit et al. (2008)

[1] Frequently used within environmental studies

ducted to date to have the most potential to track sediment-associated trace metals in rivers. It should be noted, however, that the potential application of other non-traditional isotopes as tracers in riverine systems is currently being explored, including Ag, Cr, Fe, Mo, Ni, Sb, Se, and U (see, for example, EnvironMetal Isotopes Conference Program, 2013). Which, if any, of these isotope systems can served as effective sediment tracers is yet to be determined; thus, they will not be examined herein. In addition, we have elected not to include a discussion on the use of the stable isotopes of C, H, O, N, S as tracers as they have been extensively discussed in other sources.

5.2 Zn and Cd Isotopes

Cadmium and Zn sulfide minerals are commonly associated with one another in ore deposits. As a result, Cd is often recovered along with Zn and Pb during the processing of Zn and Pb ores (Rehkämper et al. 2011). Zn and Cd also exhibit a number of other geochemical similarities (Bullen 2011; Rehkämper et al. 2011). Both, for example, are stable at the Earth's surface in a single (+2) oxidation state, and therefore are not significantly influenced by redox processes. They also are isotopically fractionated by a number of similar processes, such as by evaporation, condensation, or electroplating.

As was the case of Nd, the isotopic abundance of the non-CHONS including Cd and Zn are expressed per 1,000 (mil, δ) or per 10,000 (ε) relative to a standard, due to extremely small variations that occur in nature. Mathematically,

$$\delta^{ij} = \left(\frac{isotopic\ ratio\ of\ sample}{isotopic\ ratio\ of\ standard} - 1 \right) \times 1,000 \tag{5.1}$$

and

$$\varepsilon^{ij} = \left(\frac{isotopic\ ratio\ of\ sample}{isotopic\ ratio\ of\ standard} - 1 \right) \times 10,000 \tag{5.2}$$

where δ^{ij} and ε^{ij} are the δ and ε values of the element, respectively, for the isotopic ratio represented by ij. In most instances, the lighter isotope is used as the denominator, in which case $\delta > 0$ is heavier than the standard and $\delta < 0$ lighter. In the case of Zn, which has five stable isotopes, most studies have utilized $^{66}Zn/^{64}Zn$ ratios (and often reported as $\delta^{66}Zn$) due to the high relative abundances of ^{66}Zn and ^{64}Zn, and the ability to precisely measure their contents in geological materials. Unfortunately, studies to date have utilized different materials as a standard, making the direct comparison of reported $\delta^{66}Zn$ values difficult (Cloquet et al. 2008). Cd isotopic abundances may be reported in terms of either δ or ε. In terms of isotopic ratios, most environmental studies have utilized $^{114}Cd/^{110}Cd$ ratios, although $^{112}Cd/^{110}Cd$ has also been used (e.g., Schmitt et al. 2009). As is the case for Zn, previous Cd isotope investigations have utilized differing standards (see Rehkämper et al. 2011 for a more detailed discussion on Cd isotopes and isotopic analysis).

Unlike many other trace metals, Zn is not a highly toxic element, but there is considerable interest in the use of Zn isotopes as a tracer because (1) they may be applied to determine the source of Zn from a wide range of materials including mining and refining products and wastes, steel processing plants, coal-fired power plants, vehicles, urban waste incinerators, car tires, and other constituents in urban runoff (Chen et al. 2008), and (2) Zn is often associated with Cd and Pb in natural and anthropogenic materials, and therefore Zn isotopes also may be of use in determining the provenance of these trace metals. It is also related to the fact that they may provide information on contaminant sources when other more commonly used isotopic systems fail (as detailed below).

5.2.1 Use of Zn Isotopes as Contaminated Sediment Tracers

To date the use of Zn isotopes as a contaminant tracer in near surface environments has been limited, particularly within riverine systems. Nonetheless, recent studies (e.g., Sivry et al. 2008; Cloquet et al. 2008; Chen et al. 2008; Bird 2011; Aranda et al. 2012) suggest that Zn isotopes hold considerable promise because the isotopic composition of specific anthropogenically produced materials can, at least in some cases, be distinguished from that found naturally in rocks, sediments, soils, etc. The ability to differentiate between natural and anthropogenic sources of Zn is aided by relatively small variations in δ^{66}Zn values ($\sim 2\,‰$) in ore, sediments, and biota (Maréchal et al. 1999, 2000; Maréchal and Albaréde 2002; Cloquet et al. 2008), and the measurable fractionation of Zn by various physical and biological processes, including those used in industry. Industrial Zn fractionation is dominated by the mass-dependent processes of evaporation, condensation, and electroplating. During ore smelting, for example, the evaporation of Zn and Cd is likely to release isotopically light isotopes in the exhaust, whereas isotopically heavy Zn and Cd will remain within the smelting residue (Mattielli et al. 2009; Sivry et al. 2008). Similarly, electroplating may result in relatively light electroplated Zn (and presumably Cd) in comparison to the parent solution (Bullen 2011; Kavner et al. 2008). Once released into the surrounding environment, fractionation may occur at low temperatures by a range of processes including biogenic uptake by micro-organisms and other biota (Stenberg et al. 2004; Weiss et al. 2005), diffusion (Rodushkin et al. 2004), adsorption onto inorganic and organic surfaces (Weiss et al. 2005; Pokrovsky et al. 2005), ion exchange (Maréchal and Albaréde 2002), and mineralization (Mason et al. 2005; Wilkinson et al. 2005) (Table 5.1).

Much of the work on Zn isotopes to date has focused on sourcing Zn from atmospheric sources (Luck et al. 1999; Sonke et al. 2002, 2008; Cloquet et al. 2006; Dolgopolova et al. 2006; Weiss et al. 2007; Gioia et al. 2008; Mattielli et al. 2009; Bigalk et al. 2010a; Thapalia et al. 2010; Juillot et al. 2011). Thapalia et al. (2010), for example, examined changes in the flux and sources of Zn to Lake Ballenger located in a highly urbanized area near Seattle, Washington. Of primary interest was the atmospheric deposition of Zn from a smelter located approximately 53 km upwind from the lake that operated from 1890 to 1985. Zn deposition from the smelter was thought to be limited, however, until 1917 when a 172 m exhaust stack was added to the facility. The overall approach was to collect, date, and analyze sediments within a core from the lake, which dated back to about 1,450 YBP. Thapalia et al. (2010) found that the core could be subdivided into 4 time periods on the basis of metal concentrations and mass sediment/metal (Zn, Cu, As, and Pb) accumulation rates (Fig. 5.1). These 4 time intervals included (1) a pre-smelting period (pre-dating 1917), (2) a period of smelter operation prior to extensive urbanization within the catchment feeding the lake (~1917–1945), (3) a period characterized by smelting and rapid urbanization within the catchment (~1945–1985), and (4) a period following the closure of the smelter and characterized by relatively stable urban land use (~1985–2007). Isotopically, $\delta^{66}Zn$ values (reported using the batch JMC 3-0749-L standard) varied by 0.50 ‰ over the length of the core. More importantly, the variations systematically correlated with the timing of smelter operation (pre-smelter, smelter and post-smelter periods) (Fig. 5.1). The $\delta^{66}Zn$ composition of the

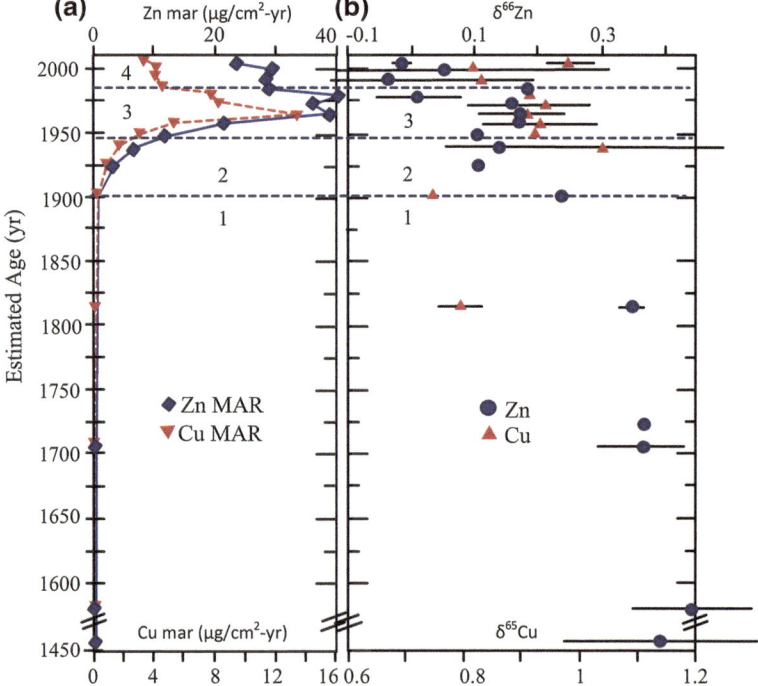

Fig. 5.1 a Mass accumulation rates (MAR) for Cu and Zn shown for core samples extracted from Lake Ballinger near Seattle, Washington, USA; **b** Zn and Cu isotopes data plotted as a function of depth within the cores (i.e., age). *Dashed blue lines* refer to the boundaries between zones *1* (presmelter period), *2* (smelter period), *3* (smelter plus urbanization period), and *4* (post smelter period). *2 error bars* denote precision of external replicates (Reprinted with permission from Thapalia et al. (2010) (Copyright 2010 American Chemical Society)

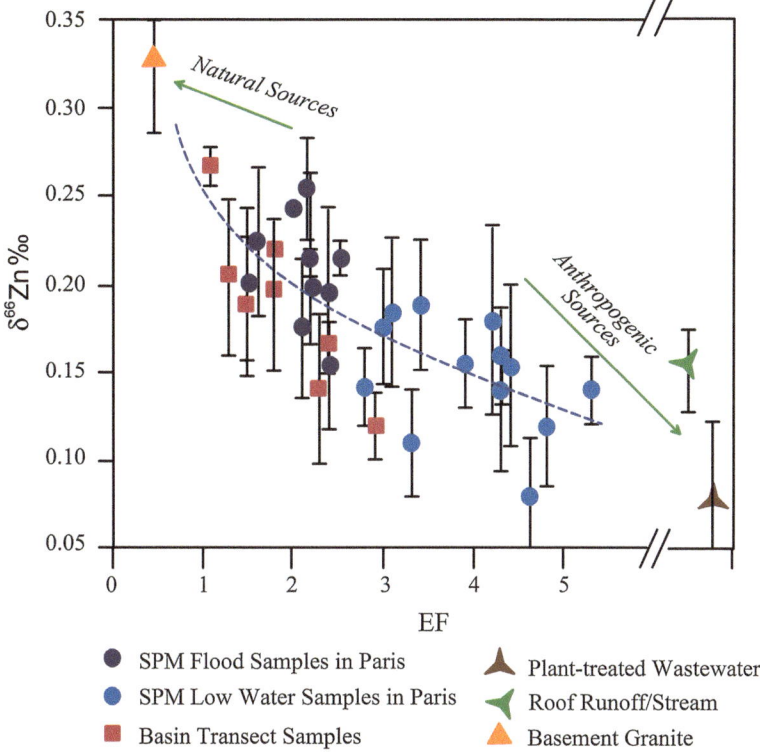

Fig. 5.2 Relation between δ^{66}Zn data and calculated enrichment factors (EF) for selected types of environmental samples from the Seine River in and near Paris, France. Samples of suspended particulate matter collected during low (*blue circles*) and flood flows (*purple circles*) and bed sediment collected along a basin transect (*red squares*) plot along a mixing line composed of two end members. End members are represented by granitic basement rocks and an anthropogenic source composed of plant-treated wastewater and roof runoff (roof stream) (modified from Chen et al. 2009)

pre-smelter sediments was thought to correspond to that of the local geological materials, whereas that of the sediments corresponding to the period following smelter operations appeared to be related to Zn in urban runoff, particularly Zn associated with the wear of vehicle tires. Interestingly, δ^{66}Zn values did not change during the period of rapid urbanization and declining smelter operations characterized by peak mass metal accumulation ($\sim$1945–1985). The limited variations in Zn isotopic values during this period were attributed to the remobilization of Zn enriched sediments from soils that had been contaminated by smelter exhausts. A rapid shift in δ^{66}Zn values did, however, occur between 1979 and 1985 (Fig. 5.1) such that the sediment acquired a signature similar to that observed during the period of smelting. Thapalia et al. (2010) argued that the alteration in δ^{66}Zn values was associated with the introduction of Zn from urban runoff that had been stored in sediments from Hall Creek that were eroded and transported to the lake prior to or during a remediation project.

One of the first detailed analyses pertaining to riverine systems was carried out by Chen and his colleagues on both the dissolved (Chen et al. 2008) and particulate (Chen et al. 2009) load within the Seine River of France. The Seine River is highly contaminated by a number of toxic trace metals (e.g., Cu, Ni, Pb, and Zn) derived from both industrial and urban sources in and around Paris. With regard to particulate matter, the study by Chen et al. (2009) consisted of two primary components: (1) a longitudinal study of the spatial changes in Zn concentrations and δ^{66}Zn values that includes reaches located both up- and downstream of Paris, and (2) a study of the changes in Zn concentration and δ^{66}Zn values through time at a site near the center of Paris. The latter primarily focused on the geochemical differences observed between low and high flow events. δ^{66}Zn values were calculated using the JMC 3-0749-L standard solution. Zn concentrations were primarily presented in terms of the magnitude of anthropogenic Zn enrichment above background values where background was taken as the Zn and Al concentrations measured in uncontaminated pre-historic deposits and forest soils.

Along the length of the channel, EFs varied semi-systematically from up- to downstream, ranging from about 1 to 5.3, respectively (concentrations ranged from about 100 to 400 ppm Zn). The increase presumably reflected the increase in Zn inputs from industrial and urban sources around Paris. δ^{66}Zn values decreased downstream from 0.30 to 0.08‰ in SPM. Temporally, the δ^{66}Zn data collected in Paris for varying flow conditions ranged from about 0.08‰ to 0.26‰. The combined spatial and temporal data exhibited an inverse relationship between EFs and δ^{66}Zn values (Fig. 5.2). Chen et al. (2009), after ruling out other factors such as adsorption, argued that the semi-systematic trends between increasing Zn concentration and decreasing δ^{66}Zn values was the result of particulate mixing of Zn from differing sources. Thus, the trend could be interpreted as a mixing curve defined by two end-members. The end-member characterized by the highest δ^{66}Zn values (lowest Zn concentrations) was consistent with a δ^{66}Zn value of 0.33‰ measured in granitic basement rocks within the catchment, and which is similar to the mean integrated value of 0.30% reported for Zn in other earth materials (Cloquet et al. 2008). The lower δ^{66}Zn end-member was interpreted to represent Zn in the particulate matter of a wastewater treatment plant (0.08–0.15‰) and samples of roof and road runoff (−0.10 to 0.08‰).

By assuming that the variations in EF and δ^{66}Zn values formed a mixing line, Chen et al. (2009) were able to determine the proportions of Zn from both natural and anthropogenic sources. They found that the proportion of anthropogenic Zn increased downstream, reaching a maximum value at the catchment mouth (pre-estuary) of 86 %; the average basin value was 62 %. Given a measured SPM Zn load of ∼315 t/yr, the Zn load from natural sources is about 44 t/yr, a value which was similar to the 42 t/yr value determined from monitoring data on the Seine River. Temporal variations in anthropogenic Zn contributions were also observed at the monitoring site in Paris. Here, anthropogenic contributions decreased within increasing discharge, and ranged from about 40 % during high flows to 100 % during low flows.

Interestingly, Zn isotopic data collected in the Seine River suggest that the source of Zn in dissolved and particulate load may differ. More specifically, δ^{66}Zn values suggested that the primary natural source of Zn within the SPM was granitic rocks (with a signature of about 0.33 ‰), whereas the primary source of dissolved Zn was Cretaceous chalk (exhibiting a δ^{66}Zn value of 0.90 ‰). Similarly, Chen et al. (2009) suggested that Zn associated with urban runoff (particularly roof runoff) adds very little particulate-associated Zn to the river, but is the primary anthropogenic source of dissolved Zn.

Sivry et al. (2008) examined Zn contamination within the Lot River basin, France, and found that Zn isotopes may also serve as effective tracers of Zn from industrial ore processing within alluvial sediments. In this case, Zn isotopes within the ore (exhibiting δ^{66}Zn ~ 0.16 ‰) was highly fractionated during metallurgical processing forming tailings materials with measurably higher δ^{66}Zn values (up to $+1.49$ ‰). In addition, they were able to demonstrate that δ^{66}Zn values within sediments of a contaminated tributary (the Riou Mort) was significantly different than the isotopic values measured in alluvial sediments located upstream of contaminant influx and assumed to represent background materials (δ^{66}Zn values of $+0.91 \pm 0.04$ ‰ as compared to $+0.31 \pm 0.06$ ‰). The observed spatial changes in δ^{66}Zn values along the river (i.e., from up- to downstream of the tributary) were consistent with the influx of tailings materials via the Riou Mort.

Sivry et al. (2008) also found that systematic changes in δ^{66}Zn values occurred within a dated sediment core extracted from a downstream hydroelectric reservoir (Fig. 5.3). Sediments deposited between approximately 1952 and 1972 exhibited a mean δ^{66}Zn value of $+0.95 \pm 0.08$ ‰. Sediments deposited during the late 1970s exhibit increasing δ^{66}Zn values until reaching a maximum value in 1986, after which they remain relatively constant until the mid-1990s (Fig. 5.3). The temporal shift to

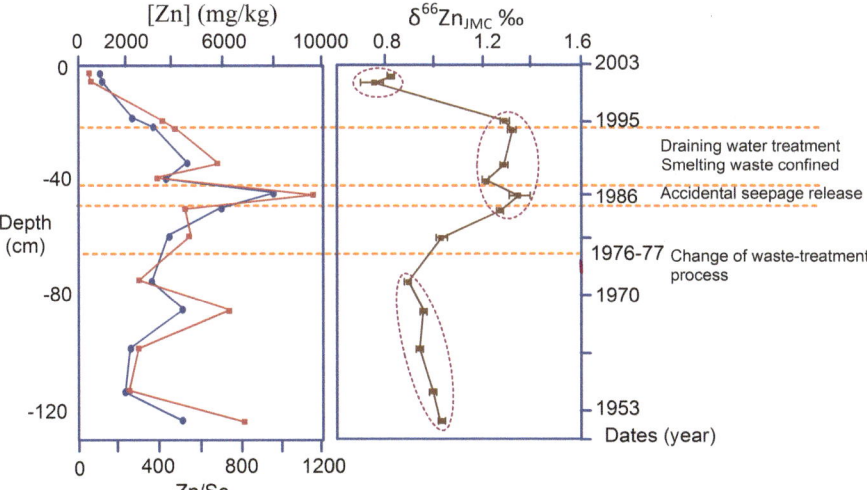

Fig. 5.3 Vertical (depth/time) variations in Zn/Sc ratios (*blue circles*), total Zn concentrations (*red squares*), and δ^{66}Zn values (*brown diamonds*) measured in samples from a core extracted from the Cajarc reservoir along the Lot River, France (from Sivry et al. 2008); Zn and Zn/Sc variations from Audry et al. (2004). The co-variation between Zn concentration and Zn isotopic data suggest that Zn isotopes can be used to determine the source of the Zn

heavier δ^{66}Zn values corresponded with a change in ore processing techniques and an increase in Zn extraction efficiencies. Thus, it appears that sediments within the reservoir were not only able to differentiate between natural and anthropogenic Zn, but Zn associated with different industrial processing methods.

Two inherent assumption in the use of both elemental concentration and isotopic data for the construction of pollution histories are that (1) the concentration or isotopic signature within the deposits correlate with the influx of metals to the river, and (2) spatial changes in concentrations or isotopic abundances reflect the variations within the river sediment at the time of deposition (Miller and Orbock Miller 2007). In other words, the post-depositional migration of trace metals within the sediment profile is limited. Several studies (e.g., Bradley and Cox 1987; Hudson-Edwards et al. 1998) have shown that migration on the order of a few centimeters can occur. In addition, isotopic abundances of many of the non-traditional isotopes, including Zn, may be affected by post-depositional (diagenetic) fractionation processes. Weiss et al. (2007), for example, found that diffusion of dissolved Zn in ombrotrophic peats altered the isotopic composition of the sediment-associated Zn, resulting in higher δ^{66}Zn values at the bottom of collected cores. However, the degree of fractionation (and mobility) depends on the physiochemical conditions at the site, and the systematic variations in isotopic ratios observed by Sivry et al. (2008) (and Thapalia et al. 2010) suggest that in the case of Zn such post-depositional alterations may not be so significant that they inhibit the use of Zn isotopes to decipher changes in Zn provenance through time or space.

5.2.2 Use of Cd Isotopes as Contaminated Sediment Tracers

In comparison to Zn, exploratory investigations of the use of Cd isotopes as an environmental tracer are more limited. In fact, with the exception of a few preliminary studies, their application to contaminated sediments in rivers is completely lacking. Thus, while the few studies that have been conducted suggest that Cd isotopes may serve as effective tracers to determine Cd provenance, their actual use has yet to be effectively demonstrated.

In a detailed review of the existing Cd isotope data, Rehkämper et al. (2011) found that Cd isotopic composition of igneous rocks derived from the crust and mantle exhibited $\varepsilon^{114/110}$Cd values between -4 and 4. Loess and clastic sediments from rivers and continental margins also fell within this narrow range of values (Fig. 5.4a). Thus, they concluded that there was no "evidence for systematic Cd isotope fractionation between or within the silicate Earth." In contrast, Cd isotope values of extraterrestrial, impact, marine and polluted materials exhibited a wider range of values. With regards to the latter, most attention to date has focused on the Cd isotope fractionation associated within the processing of Pb and Zn ores with which Cd is associated. Cloquet et al. (2006), for example, analyzed the Cd isotopic composition of dust trapped in exhaust filters and slag produced in the furnace of a Pb-Zn refinery. They found that dust exhibited lighter ε^{114}Cd values than the slag by about -10ε.

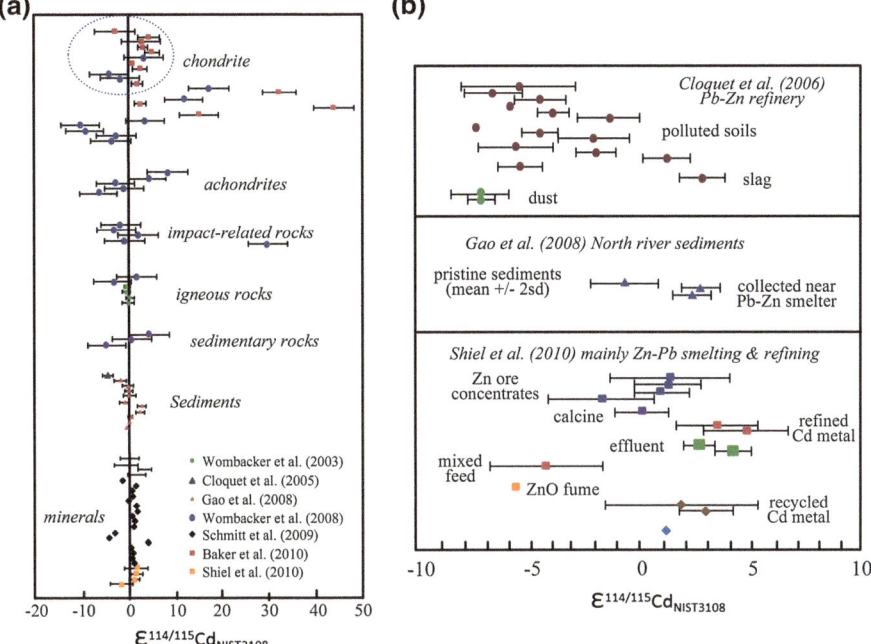

Fig. 5.4 **a** Cd isotope data for various Earth materials and meteorites. The sampled meteorites are limited to enstatite chondrites (of type EH), achondrites (eucrites and SNC meteorites from Mars), and carbonaceous chondrites. **b** Cd Isotope data obtained on samples from Zn-Pb smelting/refining plants and various types of material contaminated by smelter emissions. All samples reported relative to the NIST 3108 Cd standard. *Circled data* represents results from unfractionated carbonaceous chondrites that have not been overprinted by strong metamorphism (Wombacher et al. 2003, 2008). Shown uncertainties are reported as shown in original publications, except for the average calculated for 5 unpolluted river sediments (Gao et al. 2008) (figure from Rehkämper et al. 2011)

Like Zn, the observed difference was interpreted to result from evaporation processes. In addition, they found that analyzed soil materials located up to 4.5 km from the refinery exhibited systematic spatial variations in ε^{114}Cd values, with the lowest values occurring closest to the smelters. The observed variations were attributed to the mixing of Cd in the furnace/exhaust dust within slag-derived materials. It appeared, then, that Cd isotopes served as an effective tracer of anthropogenic Cd, at least over a relatively short distance. The argument is supported by the finding of (1) Shiel et al. (2010) who studied the Cd isotopic composition of materials associated with a Pb-Zn smelter and refining plant, this one located in Canada, and (2) Shiel et al. (2012) who were able to use a combination of Pb, Zn, and Cd isotopes to determine the extent of anthropogenic Cd accumulation in marine bivalues off the coast of British Columbia and the Eastern US.

With regards to riverine environments, we know of only one study using Cd isotopes as a tracer of anthropogenic Cd; that of Gao et al. (2008) who worked on the North River of southern China. The North River is highly contaminated, in part, by mining and smelting operations. In fact, Cd enrichment factors calculated for samples collected along the river and its tributaries are extremely high, ranging from 11.5 to 1,100 (Gao et al. 2013). Gao et al. (2008) showed that two samples collected near a Pb-Zn smelter exhibited ε^{114}Cd values that were distinctly different from uncontaminated sediments (Fig. 5.4b). The observed differences led them to suggest that Cd isotopes may serve as an effective tracer. However, additional data provided on samples collected along the North River and downstream of the smelters were less convincing. The range of ε^{114}Cd variations (relative to SPEX Cd reference solution) was limited to 0.42ε (-0.35–0.07). In addition, 10 of the samples, while exhibiting markedly different ε^{114}Cd values in comparison to effluent from the smelters, were similar, falling within a narrow range, in spite of differences in Cd concentration within the samples or the location at which the samples were located with respect to the upstream smelters. Gao et al. (2013) argued that the differences in ε^{114}Cd values that were observed were likely to reflect the mixing of dust from the smelter, slag from the smelter, and local background/ore materials.

5.3 Copper Isotopes

Copper is an essential (nutrient) trace metal, but is highly toxic to aquatic, photosynthetic microorganisms and algae. It can also be toxic to higher trophic level animals in which it can cause a condition known as oxidative stress. Its release, then, to aquatic environments is often of considerable concern. Unlike Cd and Zn, Cu is stable in the near surface environment in two oxidation states, Cu^+ (CuI) and Cu_2^+ (CuII). It therefore participates in a number of abiotic and biotic redox reactions. CuI is the common form associated with sulfide minerals including chalcopyrite ($CuFeS_2$), chalcocite (Cu_2S), enargite (Cu_3AsS_4), and covellite (CuS_2), whereas CuII is the common aqueous form. Cu may also occur on rare occasions as a native element. It possesses two isotopes, ^{65}Cu and ^{63}Cu, with relative abundances of 30.83 % and 69.17 %, respectively. Like most non-CHONS, the Cu isotopic composition of physical and biological materials is reported in units of per mil (δ) typically (but not always) relative to the NIST 976 Cu standard.

   Analyses of the Cu isotopic composition of Earth and biological materials are still limited, particularly for silicate rocks and sediments. Data collected to date on dust, deep sea sediments, sandstones, shale and basalts show that δ^{65}Cu values fall within a relatively narrow range ($+0.16 \pm 0.16$‰) (Vance et al. 2008, citing data from Maréchal et al. 1999; Archer and Vance 2004; Asael et al. 2007). These values are similar to those often reported for natural (uncontaminated) soils, which exhibit variations of ~ 1‰ with most around 0‰ (Bigalke et al. 2009). In marked contrast, the δ^{65}Cu values measured in ores and minerals are highly variable, ranging from -17 to $+10$‰ (Mathur et al. 2009). The extremes of δ^{65}Cu values noted for Cu ores and minerals appear to be rather uncommon, prompting some investigators (e.g., Weiss et al. 2008) to argue that δ^{65}Cu values typically fall within a range of about -3 to $+5.7$‰ for sediment, secondary ore minerals, and biological materials, the larger variations generally associated with low temperature secondary minerals (Hoefs 2009).

Cu also differs from Cd and Zn in that it (Cu) does not appear to be significantly fractionated by smelting processes, possibly because it possesses a boiling temperature above that which is generally reached during smelting (Gale et al. 1999; Mattielli et al. 2006). Thus, the products of smelting (exhaust, dust, slag, etc.) will reflect the δ^{65}Cu values inherent in the processed ore. The large range of δ^{65}Cu values that have been measured in ore deposits, and other Earth and biological materials, appears to be primarily related to secondary mineral forming processes (Asael et al. 2007; Bigalke et al. 2010b).

Redox reactions appear to be particularly important drivers of Cu fractionation. For example, a number of studies have shown that the abiotic oxidative dissolution of Cu sulfide minerals, in which CuI is the common form, will result in an aqueous (CuII) product being heavier than the original mineral (Mathur et al. 2005; Fernandez and Borrok 2009; Kimball et al. 2009; Maher et al. 2011; Wall et al. 2011). The oxidative dissolution of Cu sulfide minerals in the presence of certain microbes can also cause fractionation; however, the magnitude of microbial-mediated fractionation appears to decrease (Mathur et al. 2005; Kimball et al. 2009; Rodríguez et al. 2013). Fractionation has also been observed during reduction in which case CuII in solution forms a precipitated sulfide. In this case, the mineral tends to be enriched in ^{65}Cu, making the solution isotopically heavier than the mineral. Ehrlich et al. (2004), for example, found that the formation of covellite (CuI) from a solution containing CuII under anoxic conditions led to δ^{65}Cu values that were $\sim$3 % heavier in the solution. Biotic reduction of CuII to CuI has also been shown to induce fractionation by the incorporation of ^{63}Cu into the cell (Navarrete et al. 2011). Some industrial processes involving chemical reduction, such as associated with electroplating, may also lead to strong Cu fractionation, with the deposited Cu metal being isotopically lighter than the remaining Cu in the solution (Bigalk et al. 2010a).

Other potentially important Cu fractionation processes include (1) sorption/ adsorption onto mineral surfaces (Balistrieri et al. 2008; Pokrovsky et al. 2008; Vance et al. 2008) and microbes (Pokrovsky et al. 2008; Bigalke et al. 2010b, c), (2) the incorporation of Cu into bacterial cells and proteins (Zhu et al. 2002; Navarrete et al. 2011), and (3) the preferential uptake of ^{63}Cu in plants (Rodríguez et al. 2013).

The large scale fractionation of Cu by redox processes has led some investigators to argue that Cu isotopes may be an effective tracer to assess that nature of various redox reactions in natural systems (Hoefs 2009). For example, the abiotic and biotic oxidation of Cu sulfide minerals is a common process along many mine contaminated river systems resulting in acid mine drainage. As a result, variations in the Cu isotopic values observed between sulfide minerals, river waters, and secondary mineral precipitates may be useful for assessing various biogeochemical processes that influence the dispersal and cycling of Cu within mine contaminated riverine environments (Borrok et al. 2009; Fernandez and Borrok 2009; Bird 2011).

While Cu fractionation by low temperature processes may be beneficial for assessing Cu biogeochemical cycling, particularly with regards to redox reactions, some have argued that these processes may limit the use of Cu isotopes as an effective tracer of Cu pollution (Hoefs 2009). For example, within the sedimentary components of aquatic environments, their use may be complicated by redox reactions that alter the Cu isotopic ratios both during dispersal and after deposition. In addition, Cu isotopic variations with natural materials are likely to overlap with those found within anthropogenic Cu sources (Hoefs 2009; Thapalia et al. 2010).

In spite of the recognized difficulties of using Cu isotopes as an effective tracer of Cu provenance, a few recent studies have shown that their use will depend on the local environmental conditions and the nature of the Cu isotopic composition of the background and anthropogenic sources. Thapalia et al. (2010), for example, found that Cu isotopic values exhibited similar trends to those observed for Zn within the Lake Ballinger core discussed above. More specifically, measurable shifts in δ^{65}Cu values correlated to the operational history of the smelter located upwind of the site (Fig. 5.1). In contrast to Zn isotopes, however, δ^{65}Cu values in sediments that pre-dated smelting activity were lighter than those that were deposited during and following smelting operations. Thapalia et al. (2010) also suggested that Cu isotopic data may be effective for determining the source of Cu from urban sources, such as from tires and automobile emissions (although perhaps not as effective as Zn isotopes).

Using a rather novel approach, El Azzi et al. (2013) demonstrated that Cu isotopes may also be effective at determining the source and fate of Cu in riverine environments. Their study focused on a small Mediterranean catchment in southern France (the Baillaury catchment) where Cu-based pesticides (referred to as the Bordeaux mixture) were applied to vineyards to protect grapevines from fungus. Within the catchment, they sampled actively cultivated soils and abandon soils, both of which served as potential sources of anthropogenic Cu to the river. They also sampled channel bed sediments, suspended particulate matter (SPM), and river water during a flood event in February, 2009. They found that Cu enrichment factors were generally greater than 2 and, based on Cu concentrations, that anthropogenic Cu contributions ranged between 50 % in the abandoned soils to more than 75 % in the cultivated soils, river bed sediments, and SPM. Isotopic analyses performed on the bulk samples found that the local bedrock (+0.07 %), river bed sediment (−0.10 to +0.06), and SPM (+0.09) exhibited similar δ^{65}Cu values, but differed from the water samples (+0.31) and analyzed soils (−0.06 to −0.14). Based on these bulk analyses, it appears that Cu isotopes could not be used as a tracer of anthropogenic Cu. However, El Azzi et al. (2013) recognized that natural Cu is often derived from rock forming minerals and where it is bound to silicate and oxide minerals is relatively immobile. Cu associated with these minerals is usually found to be associated with the residual fraction when a sequential extraction method is applied to the sediments. In contrast, anthropogenic Cu is mostly associated with various non-residual phases and, therefore, is more mobile. Thus, they reasoned that the isotopic analyses of sequentially extracted phases could be used to quantify the magnitude of natural and anthropogenic sources. In doing so, they found using sequential extraction methods that the residual phase accounted for more than 60 % of the Cu in soils of the abandoned vineyard, but only about 30 % in the cultivated soils, river bed sediments, and SPM. Cu associated with the residual phase exhibited δ^{65}Cu values similar to that of the bedrock, and could therefore be interpreted as coming from natural sources. Cu associated with the surface horizons of the contaminated soils, and which was mostly bound to the organic fraction, exhibited a signature similar to the Bordeaux mixture used as a fungicide treatment (with or without a minor amount of fractionation). Interestingly, Cu associated with the water soluble fraction of the SPM possessed δ^{65}Cu values similar to Cu dissolved in the water. Thus, most of the Cu associated with the SPM was interpreted to have been derived from the water. The investigation is important in that it demonstrates that Cu isotopes may not only be used to gain insights into redox processes within an aquatic environment, but Cu sources and dispersal pathways.

5.4 Mercury Isotopes

Hg, a priority global pollutant, exhibits a highly complex biogeochemical cycle, a feature attributed to the fact that it possesses a stable gaseous form (Hg^0) and may undergo a range of redox, phase, and biologically mediated transformations. In fact, methylmercury, the chemical form predominantly found in biota, is primarily produced by sulfate reducing microbes. Until recently, understanding the sources and biogeochemical cycling of Hg in the near surface environment relied on an analysis of total Hg concentrations within the media of interest, its chemical speciation, and/or the application of mass balance and geochemical models (Sonke et al. 2010). However, with the recent ability to analyzed for Hg isotopes in abiotic and biotic materials it has become clear that numerous inorganic and organic reactions result in Hg isotopic fractionation (Table 5.1), producing large variations in the Hg isotopic composition of both natural and anthropogenic materials (Fig. 5.5) (Bergquist and Blum 2009). These variations in Hg isotopic values are now being used to determine Hg source(s), pollution histories, and transformations during biogeochemical cycling (Bergquist and Blum 2009; Yin et al. 2010; Sonke et al. 2010; Mil-Homens et al. 2013; Sherman and Blum 2013). In fact, it could be argued on the basis of the number of publications on the topic that more effort has been devoted to Hg isotope systematics for environmental studies than for any other of the non-CHONS.

Hg possess seven stable isotopes, and unlike the majority of the elements variations in Hg isotopic values results from both mass-dependent and mass-independent processes (Table 5.1). Mass-independent fractionation of Hg has only been observed to affect the odd isotopes (^{199}Hg, ^{201}Hg). Thus, the ratio of ^{199}Hg to ^{201}Hg appears to be particularly useful for characterizing chemical pathways, such as photochemical reduction, associated with mass-independent fractionation processes (Bergquist and Blum 2009). Mass-dependent isotopic compositions are reported using δ notation relative to the NIST Hg standard 3133 as expressed by:

$$\delta^{202}Hg = \left(\frac{^{202}Hg/^{198}Hg_{Sample}}{^{202}Hg/^{198}Hg_{SRM3133}} - 1 \right) \times 1,000 \qquad (5.3)$$

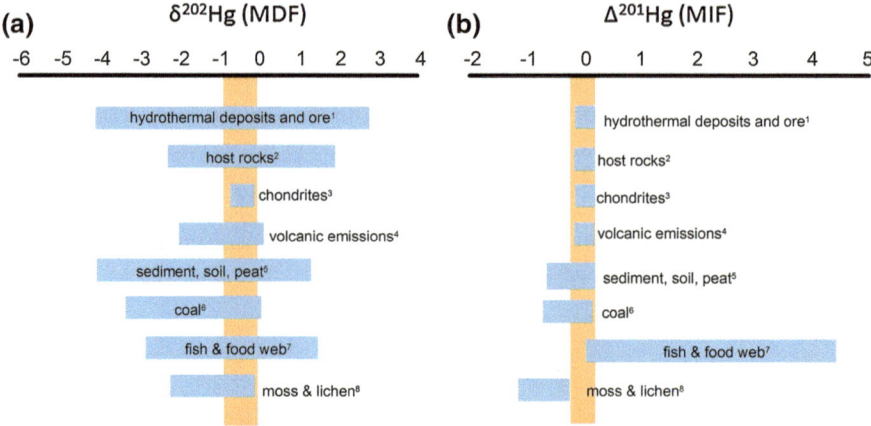

Fig. 5.5 Summary of published Hg isotopic data (from Bergquist and Blum 2009). *Horizontal bars* show range in mass-dependent fractionation (**a**) and mass-independent fractionation (**b**). The *vertical bars* represent the estimated crust- and mantel-derived Hg. Data sources include: 1. Hintelmann and Lu (2003), Smith et al. (2005, 2008); 2. Smith et al. (2008); 3. Lauretta et al. (2001); 4. Sonke et al. (2008), Zambardi et al. (2009); 5. Jackson et al. (2004), Foucher and Hintelmann (2006), Biswas et al. (2008), Ghosh et al. (2008), Jackson et al. (2008), Foucher et al. (2009), Gehrke et al. (2009); 6. Biswas et al. (2008), Ghosh et al. (2008), Carignan et al. (2009), Bergquist and Blum (2009)

Mass-independent fractionation of the odd isotopes is reported as the difference between the measured δ^{199}Hg and δ^{201}Hg values to the predicted mass-dependent values as follows (Blum and Bergquist 2007):

$$\Delta^{199}\text{Hg} = \delta^{199}\text{Hg} - (\delta^{202}\text{Hg} \cdot 0.252) \tag{5.4}$$

and

$$\Delta^{201}\text{Hg} = \delta^{201}\text{Hg} - (\delta^{202}\text{Hg} \cdot 0.752) \tag{5.5}$$

Studies of the effects of metallurgical processes on Hg isotopes suggest that waste products will often possess a distinct Hg isotopic signature. It should come as no surprise, then, that the initial studies of Hg isotopes within samples from sedimentary deposits, including those within riverine environments, have demonstrated that Hg sources have the potential to determine both Hg provenances and loading histories (Sonke et al. 2010; Foucher et al. 2009, 2013; Feng et al. 2010; Mil-Homens et al. 2013; Yin et al. 2013). Sonke et al. (2010), for example, found that within the Lot River basin, France metal refining resulted in slag residues that exhibited heavy Hg isotopic values. As a result, both the mass-dependent and mass-independent Hg isotopic signatures of contaminated sediments were isotopically heavier than that of the local background materials. As was the case for Cd, Cu, and Zn, a primary concern in the use of Hg isotopes is whether the contaminated source sediments will retain the original signature during its dispersal through the river system or whether a wide range of fractionation processes will alter its isotopic composition in undecipherable ways (Table 5.1) (Foucher et al. 2013). The analysis by Sonke et al. (2010) suggests that Hg isotopes may behave conservatively in some sedimentary environments, and may therefore serve as a potential a tracer of Hg provenance, transport processes, loading histories, and biogeochemical processes in river systems. Although the use of Hg isotopes as a tracer of Hg source(s) is clearly complicated, other studies (e.g., Donovan et al. 2013; Foucher et al. 2013; Yin et al. 2013) have supported the conclusions of Sonke et al. (2010) that Hg has significant potential as a tracer of contaminated sediment.

5.5 Summary and Management Implications

Advances in analytical chemistry, particularly the advent of the MC-ICP-MS, have created the possibility that a wide range of non-traditional stable isotopes (non-CHONS) can potentially be used as environmental tracers. In riverine sedimentary environments, recent studies have shown that Cd, Cu, Hg, and Zn may hold particular promise. In contrast to Pb, differences in the isotopic composition of geological and anthropogenic materials are due to physical and chemical isotopic fractionation processes. As a result, these isotopes allow for a linkage between trace metal concentrations and specific industrial processes used during a given time period at a single source location. In addition, some isotopic systems (e.g., Cu which is redox sensitivity) may be used to gain insights into biogeochemical cycling within river systems. The use of these non-CHONS, however, is complicated by the potential for isotopic fractionation during contaminant dispersal and/or following deposition. Fractionation may be particularly problematic for redox sensitive elements (e.g., Cu) because significant abiotic or biotic fractionation may be associated with redox reactions that occur as the sediments are dispersed through the drainage network or following deposition. Such alterations, the degree to which will depend on the physiochemical conditions of the site, may (1) inhibit the use of these non-traditional isotopes as effective tracers, and (2) require that detailed information regarding the cycling of the utilized isotopes within the studied site be collected for their effective use.

Non-CHONS will likely be used only in a supportive role in environmental forensic investigations in the near future, at least until their effectiveness as tracers of contaminated particles have been more fully demonstrated and the processes that control the nature and magnitude of isotopic fraction more completely understood. The use of the non-CHONS will likely be driven by the desire to trace specific contaminants, or to compliment data derived from other isotopic tracers. Bigalk et al. (2010a), for example, used a combination of Cu and Zn isotopes to determine the source and dispersal of Cu and Zn in soils adjacent to a Cu smelter in Slovakia. They found the $\delta^{65}Cu$ values varied little in organic rich surface soils and smelter wastes (e.g., slag, sludge, and ash). Cu isotopes, then, provided ineffective as a tracer of Cu/Zn source. However, significant variations in $\delta^{65}Cu$ values occurred with depth in the soils,

Cu becoming isotopically lighter to a depth of about 0.4 m. They attributed the Cu isotopic variations to adsorption processes, and argued that the δ^{65}Cu values could be used to assess the depth to which smelter-derived Cu was transported downward within the soil since smeltering operations had been initiated. In contrast, Zn isotopic values varied between ash, organic horizons, bedrock and slag. Moreover, the nature of the variations was such that they could be explained by mixing processes. Zn isotopic variations within the organic rich surface layers could be explained by mixing of natural soil materials with smelter Zn and the cycling of Zn within the soil-plant system. Both processes produced isotopically ligher Zn values. Within the mineral soil (below the organic horizons), small variations followed a defined mixing line between isotopically light Zn leached from the organic horizons and background Zn in the soil. Evidence of Zn fractionation by adsorption processes were lacking. Thus, a determination of the relative contributions of the trace metals with the soils and their subsequent downward movement through the soil required the combined use of Cu and Zn isotopes. Similar studies in which multiple isotopic tracers are applied to more fully understand the temporal and spatial changes in trace metal source and source contributions are likely to become more common in the future, not only with regards to contaminated soils, but riverine environments.

References

Aranda S, Borrok DM, Wanty RB, Balistrieri LS (2012) Zinc isotope investigation of surface and pore waters in a mountain watershed impacted by acid rock drainage. Sci Total Environ 420:202–213

Archer C, Vance D (2004) Mass discrimination correction in multiple-collector plasma source mass spectrometry: an example using Cu and Zn isotopes. J Anal At Spectrom 19:656–665

Asael D, Matthews A, Bar-Matthews M, Halicz L (2007) Copper isotope fractionation in sedimentary copper mineralization (Tinma Valley, Israel). Chem Geol 243:238–254

Audry S, Schafer J, Blanc G, Jouanneau JM (2004) Fifty-year sedimentary record of heavy metal pollution (Cd, Zn, Cu, Pb) in the Lot River reservoirs (France). Environ Pollut 132:413–426

Balistrieri LS, Borrok DM, Wanty RB, Ridley WI (2008) Fractionation of Cu and Zn isotopes during adsorption onto amorphous Fe(III) oxyhydroxide: experimental mixing of acid rock drainage and ambient river water. Geochim Cosmochim Acta 72:311–328

Baskaran M (2011) Environmental isotope geochemistry: past, present, and future. In: Baskaran M (ed) Handbook of environmental isotope geochemistry. Advances in Isotope Geochemistry. Springer, Berlin, pp 3–10

Bentahila Y, Othman B, Luck JM (2008) Stontium, lead, and zinc isotopes in marine cores as tracers of sedimentary provenance: a case study around Taiwan orogen. Chem Geol 248:62–82

Bergquist B, Blum JD (2009) The odds and evens of mercury isotopes: applications of mass-dependent and mass-independent isotope fractionation. Elements 5:353–357

Bigalke M, Weyer S, Wilcke W (2009) Isotopic fractionation of copper during soil genesis. Geochim Cosmochim Acta 73:A121

Bigalk M, Weyer S, Kobza J, Wilcke W (2010a) Stable Cu and Zn isotope ratios as tracers of sources and transport of Cu and Zn in contaminated soil. Geochim Cosmochim Acta 74:6801–6813

Bigalke M, Weyer S, Wilcke W (2010b) Stable copper isotopes: a novel tool to trace copper behavior in hydromorphic soils. Soil Sci Soc Am J 74:60–73

Bigalke M, Weyer S, Wilcke W (2010c) Isotopic fractionation of Cu during complexation with insolubilized humic acid. Environ Sci Technol 44:5496–5502

Bird G (2011) Provenancing anthropogenic Pb within the fluvial environment: developments and challenges in the use of Pb isotopes. Environ Int 37:802–819

Biswas A, Blum JD, Bergquist BA, Keeler GJ, Xie Z (2008) Natural mercury isotope variation in coal deposits and organic soils. Environ Sci Technol 42:8303–8309

Borrok DM, Wanty RB, Ridley WI, Lamothe PJ, Kimball BA, Verplanck PL, Runkel RL (2009) Application of iron and zinc isotopes to track the sources and mechanisms of metal loading in a mountain watershed. Appl Geochem 24:1270–1277

Bradley SB, Cox JJ (1987) Heavy metals in the Hamps and Maniforld valleys, North Straffordshire, UK: partitioning of metals in floodplain soils. Sci Total Environ 65:135–153

Blum JD, Bergquist BA (2007) Reporting of variations in the natural isotopic composition of mercury. Anal Bioanal Chem 388:353–359

Bullen T (2013) The rise of the non-CHONS and the evolution of metal and metalloid stable isotope biogeochemistry beyond non-traditional. In: EnvironMetal isotopes EMI2013, international conference: stable metal isotope fractionation processes and applications in environmental geochemistry, Ascona Switzerland, conference program and abstract booklet, p 1

Bullen TD (2011) Stable isotopes of transition and post-transition metals as tracers in environmental studies. In: Baskaran M (ed) Handbook of environmental isotope geochemistry. Advances in Isotope Geochemistry. Springer, Berlin

Bullen TD, Walczyk T (2009) Environmental and biomedical application of natural metal stable isotope variations. Elements 5:381–385

Carignan J, Estrade N, Sonke JE, Donard OFX (2009) Odd isotope deficits in atmospheric Hg measured in Lichens. Environ Sci Technol 43:5660–5664

Chen JB, Gaillardet J, Louvat P (2008) Zinc isotopes in the Seine River waters, France: a probe of anthropogenic contamination. Environ Sci Technol 42:6494–6501

Chen JB, Gaillardet J, Louvat P, Huon S (2009) Zn isotopes in the suspended load of the Seine River, France: isotopic variations and source determination. Geochim Cosmochim Acta 73:4060–4076

Cloquet C, Carignan J, Libourel G (2006) Isotopic composition of Zn and Pb atmospheric depositions in an urban/periurban area of northeastern France. Environ Sci Technol 40:6594–6600

Cloquet C, Carignan J, Lehmann MR, Vanhaecke F (2008) Variation in the isotopic composition of zinc in the natural environment and the use of zinc isotopes in biogeosciences: a review. Anal Bioanal Chem 390:451–463

Dolgopolova A, Weiss DJ, Seltmann R, Kober B, Mason TFD, Coles B, Stanley CJ (2006) Use of isotope ratios to assess sources of Pb and Zn dispersed in the environment during mining and ore processing within the Orlokava-Spokinoe mining site (Russia). Appl Geochem 21:563–579

Donovan PM, Blum JD, Yee D, Gehrke GE, Singer MB (2013) An isotopic record of mercury in San Francisco Bay sediment. Chem Geol 349–350:87–98

Ehrlich S, Butler I, Halicz L, Rickard D, Oldroyd A, Matthews A (2004) Experimental study of the copper isotope fractionation between aqueous Cu(II) and covellite, CuS. Chem Geol 209:259–269

El Azzi D, Viers J, Guiresse M, Probst A, Aubert D, Caparros J, Charles F, Guizien K, Probst JL (2013) Origin and fate of copper in a small Mediterranean vineyard catchment: new insights from combined chemical extraction and δ^{65}Cu isotopic composition. Sci Total Environ 463–464:91–101

Feng X, Foucher D, Hintelmann H, Yan H, He T, Qiu G (2010) Tracing mercury contamination sources in sediments using mercury isotope compositions. Environ Sci Technol 44:3363–3368

Fernandez A, Borrok DM (2009) Fractionation of Cu, Fe, and Zn isotopes during the oxidative weathering of sulfide-rich rocks. Chem Geol 264:1–12

Foucher D, Hintelmann H (2006) High-precision measurement of mercury isotope ratios in sediments using cold-vapor generation multi-collector inductively coupled plasma mass spectrometry. Anal Bioanal Chem 384:1470–1478

Foucher D, Nives O, Hintelmann H (2009) Tracing mercury contamination from the Idrija mining region (Slovenia) to the Gulf of Trieste using Hg isotope ratio measurements. Environ Sci Technol 43:33–39

Foucher D, Hintelmann H, Al TA, MacQuarrie KT (2013) Mercury isotope fractionation in waters and sediments of the Murray Brook mine watershed (New Brunswick, Canada): tracing mercury contamination and transformations. Chem Geol 336:87–95

Gale NH, Woodhead AP, Stos-Gale ZA, Walder A, Bowen I (1999) Natural variations detected in the isotopic composition of copper: possible applications to archaeology and geochemistry. Int J Mass Spectrom 184:1–9

Gao B, Zhou H, Liang X, Tu X (2013) Cd isotopes as a potential source tracer of metal pollution in river sediments. Environ Pollut 181:340–343

Gao B, Liu Y, Sun K, Liang XR, Peng P, Sheng G, Fu J (2008) Precise determination of cadmium and lead isotopic compositions in river sediments. Anal Chim Acta 612:114–120

Gehrke GE, Blum JD, Meyers PA (2009) The geochemical behavior and isotopic composition of Hg in a mid-Pleistocene western Mediterranean sapropel. Geochim Cosmochim Acta 73:1651–1665

Gioia S, Weiss D, Coles B, Arnold T, Babinski M (2008) Accurate and precise zinc isotope ratio measurements in urban aerosols. Anal Chem 80:9776–9780

Ghosh S, Xu Y, Humayun M, Odom L (2008) Mass-independent fractionation of mercury isotopes in the environment. Geochem Geophys Geosyst 9

Hintelmann H, Lu SY (2003) High precision isotope ratio measurements of mercury isotopes in Cinnabar ores using multi-collector inductively coupled plasma mass spectrometry. Analyst 128:635–639

Hoefs J (2009) Stable isotope geochemistry. Springer, New York

Hudson-Edwards KA, Macklin MG, Curtis CD, Vaughan DJ (1998) Chemical remobilization of contaminated metals within floodplain sediments in an incising river system: implications for dating and chemostratigraphy. Earth Surface Proc Land 23:671–684

Jackson TA, Muir DCG, Vincent WF (2004) Historical variations in the stable isotope composition of mercury in Arctic lake sediments. Environ Sci Technol 38:2813–2821

Jackson TA, Whittle DM, Evans MS, Muir DCG (2008) Evidence for mass-independent and mass-dependent fractionation of the stable isotopes of mercury by natural processes in aquatic ecosystems. Appl Geochem 23:547–571

Juillot F, Maréchal C, Morin G, Jouvin D, Cacaly S, Telouk P, Benedetti MF, Ildefonse P, Sutton S, Guyot F, Brown GE Jr (2011) Contrasting isotopic signatures between anthropogenic and geogenic Zn and evidence for post-depositional fractionation processes in smelter-impacted soils from Northern France. Geochim Cosmochim Acta 75:2295–2308

Kavner A, John SG, Sass S, Boyle EA (2008) Redox-driven stable isotope fractionation in transition metals: application to Zn electroplating. Geochim Cosmochim Acta 72:1731–1741

Kimball BE, Mathur R, Dohnalkova AC, Wall AJ, Runkel RL, Brantley SL (2009) Copper isotope fractionation in acid mine drainage. Geochim Cosmochim Acta 73:1247–1263

Lauretta DS, Klaue B, Blum JD, Buseck PR (2001) Mercury abundances and isotopic compositions in the Murchison (CM) and Allende (CV) carbonaceous chondrites. Geochim Cosmochim Acta 65:2807–2818

Luck JM, Ben Othman D, Albarède F, Télouk P (1999) Pb, Zn and Cu isotopic variation and trace elements in rain. In: Armannsson (ed) Geochemistry of the earths surface. Balkema, pp 199–202

Maher K, Jackson S, Mountain B (2011) Experimental evaluation of the fluidmineral fractionation of Cu isotopes at 250 °C and 300 °C. Chem Geol 286:229–239

Maréchal C, Albaréde F (2002) Ion-exchange fractionation of copper and zinc isotopes. Geochim Cosmochim Acta 66:1499–1509

Maréchal C, Télouk P, Albarède F (1999) Precise analysis of copper and zinc isotopic compositions by plasma source mass spectrometry. Chem Geol 156:251–273

Maréchal CN, Nicolas E, Douchet C, Albarède F (2000) Abundance of zinc isotopes as a marine biogeochemical tracer. Geochem Geophys Geosys 1:1999GC000029

Mason TFD, Weiss DJ, Chapman JB, Wilkinson JJ, Tessalina SG, Spiro B, Horstwood MSA, Spratt J, Coles BJ (2005) Zn and Cu isotopic variability in the Alexandrinka volcanic-hosted massive sulphide (VHMS) ore deposit, Urals, Russia. Chem Geol 221:170–187

Mathur R, Ruiz J, Titley S, Liermann L, Buss H, Brantley S (2005) Cu isotopic fractionation in the supergene environment with and without bacteria. Geochim Cosmochim Acta 69:5233–5246

Mathur R, Titley S, Barra F, Brantley S, Wilson M, Phillips A, Munizaga F, Maksaev V, Vervoort J, Hart G (2009) Exploration potential of Cu isotope fractionation in porphyry copper deposits. J Geochem Explor 102:1–6

Mattielli N, Rimetz J, Petit J, Perdrix E, Deboudt K, Flament P, Weis D (2006) ZnCu isotopic study and speciation of airborne metal particles within a 5 km zone of a lead zinc smelter. Geochim Cosmochim Acta 70:A401

Mattielli N, Petit JCJ, Deboudt K, Flament P, Perdrix E, Taillez A, Rimetz-Planchon J, Weis D (2009) Zn isotope study of atmospheric emissions and dry depositions within a 5 km radius of a PbZn refinery. Atmos Environ 43:1265–1272

Mil-Homens M, Blum J, Canário J, Caetano M, Costa AM, Lebreiro SM, Trancoso M, Richter T, de Stigter H, Johnson M, Branco V, Cesrio R, Mouro F, Mateus M, Boer W, Melo Z (2013) Tracing anthropogenic Hg and Pb input using stable Hg and Pb isotope ratios in sediments of the central Portuguese Margin. Chem Geol 336:62–71

Miller JR (2013) The forensic assessment of riverine contamination by historic and modern mining activity in the 21st Century. In: Cliff D (ed) Taking health and safety in the mining industry into the 21st Century—innovative solutions to difficult problems, Special Issue, Minerals 3:192–246. doi:10.3390/min3020192

Miller JR, Orbock Miller SM (2007) Contaminated rivers: a geomorphological-geochemical approach to site assessment and remediation. Springer, Berlin

Navarrete JU, Borrok DM, Viveros M, Ellzey JT (2011) Copper isotope fractionation during surface adsorption and intracellular incorporation by bacteria. Geochim Cosmochim Acta 75:784–799

Petit JCJ, DeJong J, Chou L, Mattielli N (2008) Development of Cu and Zn isotope MC-ICP-MS measurements: application to suspended particulate matter and sediments from the Scheldt Estuary. Geostand Geoanal Res 32:149–166

Pokrovsky OS, Viers J, Freydier R (2005) Zinc stable isotope fractionation during its adsorption on oxides and hydroxides. J Colloid Interface Sci 291:192–200

Pokrovsky O, Viers J, Emnova EE, Kompantseva EI, Freydier R (2008) Copper isotope fractionation during its interaction with soil and aquatic microorganisms and metal oxy(hydr)oxides: possible structural control. Geochim Cosmochim Acta 72:1742–1757

Rehkämper M, Wombacker R, Horner TJ, Xue Z (2011) Natural and anthropogenic Cd isotope variations. In: Baskaran M (ed) Handbook of environmental isotope geochemistry. Advances in Isotope Geochemistry. Springer, Berlin, pp 125–154

Rodríguez NP, Engström EE, Rodushkin I, Nason P, Alakangas L, Öhlander B (2013) Copper and iron isotope fractionation in mine tailings at the Laver and Kristineberg mines, Northern Sweden. Appl Geochem 32:204–215

Rodushkin I, Stenberg A, Andren H, Malinovsky D, Baxter DC (2004) Isotopic fractionation during diffusion of transition metal ions in solution. Anal Chem 76:2148–2151

Schmitt AD, Galer SJG, Abouchami W (2009) Isotope fractional of cadmium in nature with emphasis on the marine environment. Earth Planet Sci Lett 277:262–272

Sherman LS, Blum JD (2013) Mercury stable isotopes in sediments and largemouth bass from Florida lakes, USA. Sci Total Environ 448:163–175

Shiel AE, Weis D, Orians KJ (2010) Evaluation of zinc, cadmium and lead isotope fractionation during smelting and refining. Sci Total Environ 408:2357–2368

Shiel AE, Weis D, Orians KJ (2012) Tracing cadmium, zinc, and lead sources in bivalves from the coasts of Western Canada and the USA using isotopes. Geochim Cosmochim Acta 76:175–190

Sivry Y, Riotte J, Sonke JE, Audry S, Schäfer J, Viers J, Blanc G, Freydier R, Dupré B (2008) Zn isotopes as tracers of anthropogenic pollution from Zn-ore smelters. The Riou MortLot River system. Chem Geol 255:295–304

Smith CN, Kesler SE, Klaue B, Blum JD (2005) Mercury isotope fractionation in fossil hydrothermal systems. Geology 33:825–828

Smith CN, Kesler SE, Blum JD, Rytuba JJ (2008) Isotope geochemistry of mercury in source rocks, mineral deposits and spring deposits of the California Coast Ranges, USA. Earth Planet Sci Lett 269:399–407

Sonke JE, Hoogewerft JA, van der Laan SR, Vangronsveld J (2002) A chemical and mineralogical reconstruction of Zn-smelter emissions in the Kempen region (Belgium), based on organic pool sediment cores. Sci Total Environ 292:101–119

Sonke JE, Sivry Y, Viers J, Freydier R, Dejonghe L, Andre L, Aggarwal JK, Fontan F, Dupré B (2008) Historical variations in the isotopic composition of atmospheric zinc deposition from a zinc smelter. Chem Geol 252:145–157

Sonke JE, Schäfer J, Chmeleff J, Audry S, Blanc G, Dupré B (2010) Sedimentary mercury stable isotope records of atmospheric and riverine pollution from two major European heavy metal refineries. Chem Geol 279:90–100

Stenberg A, Andren H, Malinovski D, Engström E, Rodushkin I, Baxter DC (2004) Isotopic variations of Zn in biological materials. Anal Chem 76:3971–3978

Thapalia A, Borrok DM, Van Metre P, Musgrove M, Landa ER (2010) Zn and Cu isotopes as tracers of anthropogenic contamination in a sediment core from an urban lake. Environ Sci Technol 44:1544–1550

Vance D, Archer C, Bermin J, Perkins J, Statham PJ, Lohan MC et al (2008) The copper isotope geochemistry of rivers and the oceans. Earth Planet Sci Lett 274:204–213

Wall AJ, Mathur R, Post JE, Heaney PJ (2011) Cu isotope fractionation during bornite dissolution: an in situ X-ray diffraction analysis. Ore Geol Rev 41:62–70

Weiss DJ, Mason TFD, Zhao FJ, Kirk GJD, Coles BJ, Horstwood MSA (2005) Isotopic discrimination of zinc in higher plants. New Phytol 165:703–710

Weiss DJ, Rausch N, Mason TFD, Coles BJ, Wilkinson JJ, Ukonmaanaho L, Arnold T, Nieminen TM (2007) Atmospheric deposition and isotope biogeochemistry of zinc in ombrotrophic peat. Geochim Cosmochim Acta 71:3498–3517

Weiss DJ, Rehkämper M, Schoenberg R, McLaughlin M, Kirby J, Campbell PGC, Arnold T, Chapman J, Peel K, Gioia S (2008) Application of nontraditional stable-isotopes to the study of sources and fate of metals in the environment. Environ Sci Technol 42:655–664

Wilkinson JJ, Weiss DJ, Mason TFD, Coles BJ (2005) Zinc isotope variation in hydrothermal systems: preliminary evidence from the Irish midlands ore field. Econ Geol 100:583–590

Wombacker F, Rehkaemper M, Mezger K, Muenker C (2003) Stable isotope composition of cadmium in geological materials and meteorites determined by multiple collector-ICPMS. Geochim Cosmochim Acta 67:4639–4654

Wombacher F, Rahkämper M, Mezger K (2004) Determination of the massdependence of cadmium isotope fractionation during evaporation. Geochim Cosmochim Acta 68:2349–2357

Wombacker F, Rehkaemper M, Mezger K, Bischoff A, Muenker C (2008) Cadmium stable isotope cosmochemistry. Geochim Cosmochim Acta 72:646–667

Yin R, Feng X, Shi W (2010) Application of the stable-isotope system to the study of sources and fate of Hg in the environment: a review. App Geochem 25:1467–1477

Yin R, Feng X, Wang J, Li P, Liu J, Zhang Y, Chen J, Zheng L, Hu T (2013) Mercury speciation and mercury isotope fractionation during ore roasting process and their implication to source identification of downstream sediment in the Wanshan mercury mining area, SW China. Chem Geol 336:72–79

Zambardi T, Sonke JE, Toutain JP, Sortino F, Shinohara H (2009) Mercury emissions and stable isotopic compositions at Vulcano Island (Italy). Earth Planet Sci Lett 277:236–243

Zhu XK, O'nions RK, Guo Y, Belshaw NS, Rickard D (2000) Determination of natural Cu-isotope variation by plasma source mass spectrometry: implications for use as geochemical tracers. Chem Geol 163:139–149

Zhu XK, Guo Y, Williams RJP, O'Nions RK, Matthews A, Belshaw NS, Canters GW, de Waal EC, Weser U, Burgess BK, Salvato B (2002) Mass fractionation processes of transition metal isotopes. Earth Planet Sci Lett 200:47–62

Appendix A
Abbreviations, Unit Conversions, and Elemental Data

See Tables A.1, A.2, A.3, A.4 and A.5.

Table A.1 List of word abbreviations (acronyms)

Abbreviation	Definition
CEC	Cation exchange capacity
DO	Dissolved oxygen
DOC	Dissolved organic carbon
FRN	Fallout radionuclide
HR-ICP-MS	High-resolution sector field inductively coupled mass spectrometer
ICP-MS	Inductively coupled mass spectrometer
MC-ICP-MS	Multi-collector, inductively coupled mass spectrometer
MORBs	Mid-oceanic ridge basalts
Non-CHONS	Stable Isotopes other than carbon, hydrogen, oxygen, nitrogen, and sulfur
SPM	Suspended particulate matter
SSC	Suspended sediment concentration
TMDL	Total maximum daily load
USLE, RUSLE	Universal soil loss equation; Revised universal soil loss equation

Table A.2 List of unit abbreviations (from Miller and Orbock Miller 2007)

English unit	Abbreviation	SI Unit	Abbreviation
Acre	ac	Hectare	Ha
Fahrenheit	F	Celsius	C
Inches	in	Centimeters	cm
Yards	yd	Meters	m
Ounces	oz	Grams	gm
Pounds	lb	Kilograms	kg
Gallons	gal	Liter	l
		Milliliter	ml

© The Author(s) 2015
J.R. Miller et al., *Application of Geochemical Tracers to Fluvial Sediment*,
SpringerBriefs in Earth Sciences, DOI 10.1007/978-3-319-13221-1

Table A.3 Units of concentration (from Miller and Orbock Miller 2007)

Concentrations	(Volume)	Concentrations	(weight)
Unit	Symbols	Unit	Symbols
Moles per liter	mol/L or M	Moles per kg	mol/kg
Millimoles per liter	mmol/L or mM	Milliequavlents per kilogram	meq/kg
Micromoles per liter	mol/L or M	Micrograms per kg or ppb	g/kg or ppb
Micrograms per liter	g/L	Milligrams per kg or parts per million	g/kg = ppm

Table A.4 Common conversions (from Miller and Orbock Miller 2007)

Distance/Length	Area
1 nm = 10^{-9} m 1 m = 10^{-6} m 1 mm = 10^{-3} m 1 cm = 10^{-2} m 1 km = 10^{3} m 1 mile = 1.609 km or 1609.34 m 1 in = 25.4 mm or 2.54 cm 1 acre = 0.4046 ha	1 ft^2 = 9.280×10^{-2} m^2 1 acre = 0.4046 ha 1 mi^2 = 2.5899 km^2
Flow/Volume	Mass/Weight
1 ml = 10^{-3} L 1 L = 10^{3} cm^3 1 gal = 3.785 L 1 quart (US) = 0.9463 L 1 ft^3 = 0.0283 m^3 1 cfs = 28.32 L/s or 0.0283 m^3/s	1 pg = 10^{-12} g 1 pg = 10^{-12} g 1 ng = 10^{-9} g 1 g = 10^{-6} g 1 mg = 10^{-9} g 1 kg = 10^{3} g 1 metric ton = 10^{3} kg 1 short ton = 2000 lbs or 907.184 kg 1 lb = 453.592 g 1 troy oz = 31.1035 g
Radioactivity	Pressure
1 rad (absorbed dose) = 10^{-2} J/kg 1 pCi = 10^{-12} Ci or 0.037 Bq 1 curie = 3.7×10^{10} dps 1 becquerel (Bq) = 1.0 dps	1 bar = 10^{5} Pa or 0.9869 atm 1 Pa = 10^{-5} bar 1 atm = 760 mm Hg or 1.01325×10^{5} Pa

Table A.5 Elemental data (from Miller and Orbock Miller 2007)

Element	Symbol	Atomic number	Atomic weight	Element	Symbol	Atomic number	Atomic weight
Hydrogen	H	1	1.0079	Helium	He	2	4.0026
Lithium	Li	3	6.941	Beryllium	Be	4	9.0122
Boron	B	5	10.811	Carbon	C	6	12.011
Nitrogen	N	7	14.007	Oxygen	O	8	15.999
Fluorine	F	9	18.998	Neon	Ne	10	20.180
Sodium	Na	11	22.980	Magnesium	Mg	12	24.305
Aluminum	Al	13	26.982	Silicon	Si	14	28.086
Phosphorus	P	15	30.974	Sulfur	S	16	32.065
Chlorine	Cl	17	35.453	Argon	Ar	18	39.948
Potassium	K	19	39.098	Calcium	Ca	20	40.078
Scandium	Sc	21	44.956	Titanium	Ti	22	47.867
Vanadium	V	23	50.942	Chromium	Cr	24	51.996
Manganese	Mn	25	54.938	Iron	Fe	26	55.845
Cobalt	Co	27	58.693	Nickel	Ni	28	58.693
Copper	Cu	29	63.546	Zinc	Zn	30	65.39
Gallium	Ga	31	69.723	Germanium	Ge	32	72.84
Arsenic	As	33	74.922	Selenium	Se	34	78.96
Bromine	Br	35	79.904	Krypton	Kr	36	83.80
Rubidium	Rb	37	85.468	Strontium	Sr	38	87.62
Yttrium	Y	39	88.906	Zirconium	Zr	40	91.224
Niobium	Nb	41	92.906	Molybdenum	Mo	42	95.94
Technetium	Te	43	98	Ruthenium	Ru	44	101.07
Rhodium	Rh	45	102.91	Palladium	Pd	46	106.42
Silver	Ag	47	107.87	Cadmium	Cd	48	112.41
Indium	In	49	114.82	Tin	Sn	50	118.71
Antimony	Sb	51	121.60	Tellurium	Te	52	127.60
Iodine	I	53	126.90	Xenon	Xe	54	131.29
Cesium	Cs	55	132.91	Barium	Ba	56	137.33
Lanthanum	La	57	137.91	Cerium	Ce	58	140.12
Praseodymium	Pr	59	140.91	Neodymium	Nd	60	144.24
Promethium	Pm	61	145	Smarium	Sm	62	150.36
Europium	Eu	63	150.96	Gadolinium	Gd	64	157.25
Terbium	Tb	65	158.93	Dysprosium	Dy	66	162.50
Holmium	Ho	67	164.93	Erbium	Er	68	167.26
Thulium	Tm	69	168.93	Ytterbium	Yb	70	173.04
Lutetium	Lu	71	174.97	Hafnium	Hf	72	178.49
Tantalum	Ta	73	180.95	Tungsten	W	74	183.84
Rhenium	Re	75	186.21	Osmium	Os	76	190.23

(continued)

Table A.5 (continued)

Element	Symbol	Atomic number	Atomic weight	Element	Symbol	Atomic number	Atomic weight
Iridium	Ir	77	192.22	Platinum	Pt	78	195.08
Gold	Au	79	196.97	Mercury	Hg	80	200.59
Thallium	Tl	81	204.38	Lead	Pb	82	207.2
Bismuth	Bi	83	208.98	Polonium	Po	84	209
Astatine	At	85	210	Radon	Rn	86	222
Francium	Fr	87	223	Radium	Ra	88	226
Actinium	Ac	89	227	Thorium	Th	90	232.94
Protactinium	Pa	91	232.04	Uranium	U	92	238.03
Neptunium	Np	93	237	Plutonium	Pu	94	244
Americium	Am	95	243	Curium	Cm	96	247
Berkelium	Bk	97	247	Californium	Cf	98	251
Einsteinium	Es	99	252	Fermium	Fm	100	257
Mendelevium	Md	101	258	Nobelium	No	102	259
Lawrencium	Lr	103	262	Rutherfordium	Rf	104	281
Dubnium	Db	105	262	Seaborgium	Sg	106	266
Bohrium	Bh	107	264	Hassium	Hs	108	277
Meitnerium	Mt	109	267	Ununnilium	Uun	110	285
Ununbium	Uub	112	285	Ununquadium	Uuq	114	289